PRIVATIZING WATER

PRIVATIZING WATER

GOVERNANCE

FAILURE

AND THE

WORLD'S URBAN

WATER CRISIS

KAREN BAKKER

Cornell University Press

ITHACA AND LONDON

First published 2010 by Cornell University Press
First printing, Cornell Paperbacks, 2010

Printed in the United States of America

Library of Congress Cataloging-in-Publication Data

Bakker, Karen J.
    Privatizing water : governance failure and the world's urban water crisis / Karen Bakker.
        p. cm.
    Includes bibliographical references and index.
    ISBN 978-0-8014-4723-5 (cloth : alk. paper) — ISBN 978-0-8014-7464-4 (pbk. : alk. paper)

    1. Municipal water supply—Political aspects.   2. Water utilities—Deregulation.   3. Privatization.   4. Right to water.   I. Title.
    HD4456.B35 2010
    363.6'1—dc22   2010015488

Cornell University Press strives to use environmentally responsible suppliers and materials to the fullest extent possible in the publishing of its books. Such materials include vegetable-based, low-VOC inks and acid-free papers that are recycled, totally chlorine-free, or partly composed of nonwood fibers. For further information, visit our website at www.cornellpress.cornell.edu.

Cloth printing   10 9 8 7 6 5 4 3 2 1
Paperback printing   10 9 8 7 6 5 4 3 2 1

CONTENTS

FIGURES

TABLES

My interest in water privatization and its links to questions of social and environmental justice began with the British drought of 1995. My arrival at the University of Oxford coincided with one of the most severe droughts Britain had experienced during the twentieth century, straining the ability of recently privatized water companies to meet demand. One private water company was particularly vulnerable to the dramatic reduction in rainfall. In an attempt to stave off water rationing and cutoffs, the company embarked on a large-scale tanker-truck operation, moving hundreds of millions of liters of water from lowland streams to its dry upland reservoirs in a round-the-clock operation—lasting several months—that journalists compared to filling an Olympic-size swimming pool with a thimble. Managers called on customers to conserve water, and even encouraged businesses to move operations out of the drought-stricken area. But customers were incensed to learn of leakage rates, and public outcry intensified when—at the height of the drought—the company's CEO was caught "sneaking" baths (despite asking consumers to refrain from bathing while water supplies were threatened).

Meanwhile, the same private water company was reporting record profits. Water prices—and company profits—had risen sharply since the privatization of the British and Welsh water supply in 1989. Company managers defended their expenditures, remuneration, and management strategies: not a single consumer had their water supply cut off during the drought. But environmentalists and consumer groups criticized the impact on local rivers and water bills, protested "fat cat" profits, and called for the industry to be taken back under public control.

In the aftermath, the British government stepped in and tightened up regulations on drought response, leakage rates, and water prices. As the drought—and the furor over the government's response—subsided, water privatization faded from the public eye. But I remained intrigued, for two reasons. First, I was curious about the impacts of privatization: How did private ownership change the management practices of water supply utilities, and with what economic, social, and environmental implications? Second, water privatization was suggestive of the political, economic, and environmental challenges that arise when governments attempt to deploy private companies to manage resources—a trend increasingly widespread by the mid-1990s. Proponents of privatization portrayed water as a final frontier: water privatization symbolized the unprecedented degree to which private companies and markets had penetrated the public domain.

The politics of protest surrounding water privatization were equally compelling. The broad-based coalitions that sprang up—uniting organized labor, consumer groups, environmental organizations, and religious groups, to name just a few—prefigured the red-green alliances that emerged a few years later at the 1999 World Trade Organization Conference (the street protests labeled by journalists as "the Battle of Seattle"). Contradictory though their viewpoints sometimes were, protestors married issues of social and ecological justice in ways that usefully paralleled academic debates about ecological governance. They provided an intriguing set of responses to the prevailing wisdom in policy and academic debates about the appropriate roles of governments and markets in managing the environment. They also evoked a set of suggestive possibilities for dealing with some of the seemingly irreconcilable tensions inherent in our relationship with nature under modernity.

Initially, I focused my research on the impact of water privatization in England and Wales on consumers and the environment. This led to a full-length book, published in 2004 as *An Uncooperative Commodity: Privatizing Water in England and Wales*. I do not refer directly to this research in the present book, although it has helped frame my thinking on privatization. In particular, it encouraged me to situate water privatization in historical and cultural context. It also reinforced my view that the political, economic, and environmental dimensions of privatization are interrelated (and that they are often inappropriately divorced in both policy and academic research).

As the private British water companies internationalized their operations in the 1990s, my research naturally followed. I grew interested in the effects of water privatization in so-called developing countries, albeit in a very different and varied set of political, economic, and ecological circumstances. I spent over a decade conducting research on the activities of private water supply companies, in Latin America, Southeast Asia, and Southern Africa. The present book synthesizes the results of this research. It should interest the wide range of individuals and organizations actively involved in the debate over water privatization in lower- and middle-income countries: bilateral aid agencies, government departments and water supply agencies, unions, private companies and the corporate "watchdogs" that monitor them, multilateral financial organizations, religious groups, environmental groups, consumer groups, and "alter-globalization" activists.

These groups are engaged in intensely fought campaigns for and against water privatization. Many, though not all, are frustrated with the stale standoff between "government" and "private" that has characterized the debate for more than a decade. My book is written with this broad audience in mind; and it is directed equally toward scholars and students with an interest in the conceptual, practical, and political possibilities opened up by a move beyond privatization—the focus of the latter half of the book.

Water privatization is subject to fierce debate, and so a comment on my personal stance is appropriate. It is impossible to remain truly neutral, although I have tried to be objective. My perspective has evolved over the past fifteen years. During my travels, I saw many poorly run government systems *and* poorly run private systems (as well as, sadly, a smaller number of well-run systems of both sorts). These experiences aroused my suspicions about simplistic arguments in favor of government or of private provision, and about generic solutions to the very complex, varied water problems often unhelpfully lumped together under the label of the global water "crisis." I do not think private provision is necessary or appropriate in many circumstances, but I do agree that our conventional approach to government provision is unsatisfactory. In part, this has arisen because of the inaccurate and imprecise definitions of what "public" and "private" actors are, and do, in supplying water.

Rethinking these terms, I decided, had to be central to an analysis of the privatization debate. And this had to be done in a concrete, specific fashion, focusing on those areas where privatization had been most advanced: the world's cities. Accordingly, I narrowed my focus to an analysis of the causes of urban water supply crises, the role of privatization, and potential responses. The analysis attempts to look "beyond privatization": to unsettle the entrenched, stale positions so often evident in the "public versus private" debate; to rethink the causes of the global water supply crisis; and to expand the terms of debates about potential responses.

I have been fortunate to receive a great deal of assistance and support over the past two decades. Numerous individuals (water company employees, consumers, government officials, private water company employees, NGO representatives) gave generously of their time. Several organizations deserve special mention: the Alternative Information and Development Center, the Council of Canadians, Friends of the Earth (Canada), the Forum on Privatization and the Public Domain, Mvula Trust, the Public Services International Research Unit, the Water Dialogues, the Municipal Services Project, the Peter Wall Institute for Advanced Studies, Thames Water, and the Urban Poor Consortium.

Michelle Kooy merits special mention and thanks as coauthor of an article on Jakarta's water supply system, aspects of which were revised and incorporated into chapter 5, of which she is also coauthor. This work drew significantly, in turn, on elements of her doctoral thesis.

Much of the research in this volume has been presented in various forms at academic conferences and workshops over the past decade. It has benefited from comments from numerous individuals (some of whom I may have inadvertently overlooked): Luis Babiano, David Barkin, Bernard Barraqué, Rutgerd Boelens, Patrick Bond, David Boys, Oliver Brandes, Gavin Bridge, David Brooks, Rocío Bustamante, Noel Castree, Gordon Clark, Alice Cohen, Christina Cook, Olivier Coutard, Adam Davidson-Harden, Leandro del Moral, Bill Dorman, Rohyn d'Souza, Gemma Dunn, Jody Emel, Melanie Feakins, Philip Fletcher, Kathryn Furlong, Mary Galvin, Matthew Gandy, Consuelo Giansante, David Hall, Leila Harris, David Hemson, Sylvy Jaglin, Jen Karmona, Roger Keil, Brewster Kneen, Tom Kruse, Nina Laurie, Andrew Leyshon, Emanuele Lobina, Alex Loftus, Bronwen Morgan, Emma Norman, Marcela Olivera, Oscar Olivera, David Lloyd Owen, Dominique Lorrain, Graciela Madanes-Schneier, Becky Mansfield, David McDonald, Leandro del Moral, Anil Naidoo, Ben Page, Jamie Peck, Laila Smith, Susan Spronk, Neil Summerton, and Erik Swyngedouw.

I would also like to acknowledge the support and encouragement of colleagues. For superb research assistance, I am grateful to Alice Cohen, Jennifer Karmona, Michelle Kooy, Ernst-Jan Martijn, Suzanne Moccia, and Nur Endah Shofiani. Eric Leinberger's expert assistance with graphics is, as always, appreciated.

I am particularly grateful for the considerable institutional support that has made this project possible. I was fortunate to have been hosted by many universities during this research, including Carleton University, the École Nationale des Ponts et Chaussées (LATTS), Queen's University, Reading University, Simon Fraser University, the School of Oriental and African Studies, the University of Oxford, University of Newcastle, University of Indonesia, Universidad de Sevilla, University of the Witwatersrand, and York University. Funding support from the Social Sciences and Humanities Research Council of Canada, the Nuffield Foundation, the British Academy, the Rhodes Trust, the University of British Columbia, the Killam Trust, the Peter Wall Institute for Advanced Studies, and the University of Oxford is gratefully acknowledged. The Department of Geography at the University of British Columbia provided a congenial environment in which this project was (re)conceived in its early days. And the Peter Wall Institute for Advanced Studies at the University of British Columbia merits special mention as the place where I did much of the writing.

Peter Wissoker at Cornell University Press recognized the potential of the manuscript; his incisive commentary was of great assistance in the early stages of its preparation. He was ably succeeded by Roger Haydon, whom I thank for seeing the revised manuscript through to final publication. David Harrison, Alice Cohen, Marie Flaherty-Jones, and Hana Boye brought editorial skills and the perspective of general-interest readers to these pages. I believe the book is stronger for their contributions.

Several chapters in the book contain or draw from material in previously published articles. Permission to reproduce this material from the following sources is gratefully acknowledged: K. Bakker, "From Archipelago to Network: Urbanization and Water Privatization in the South," *Geographical Journal* 169, no. 4 (2003): 328–341; K. Bakker, "The Commons versus the Commodity: 'Alter'-Globalization, Privatization, and the Human Right to Water in the Global South," *Antipode* 39, no. 3 (2007): 430–455; K. Bakker, "Trickle Down? Private Sector Participation and the Pro-Poor Water Supply Debate in Jakarta, Indonesia," *Geoforum* 38, no. 5 (2007): 855–868; K. Bakker, "The Ambiguity of Community: Debating Alternatives to Water Supply Privatization," *Water Alternatives* 1, no. 2 (2008): 236–252; K. Bakker, M. Kooy, E. Shofiani, and E.J. Martijn, "Governance Failure: Rethinking the Institutional Dimensions of Urban Water Supply to Poor Households," *World Development* 36, no. 10 (2008): 1891–1915; M. Kooy and K. Bakker, "Splintered Networks? Urban Water Governance in Jakarta," *Geoforum* 39, no. 6 (2008): 1843–1858; M. Kooy and K. Bakker, "Technologies of Government: Constituting Subjectivities, Spaces, and Infrastructures in Colonial and Contemporary Jakarta," *International Journal of Urban and Regional Research* 32, no. 2 (2008): 375–391.

This book is dedicated to my husband Philippe, who was there throughout the entire project, from beginning to end, as my harshest editorial critic and best friend. I need not say more, for he already knows.

ABBREVIATIONS AND ACRONYMS

ADB	Asian Development Bank
BOO	build-own-operate
BOT	build-operate/own-transfer
BPD	Building Partnerships in Development (for Water & Sanitation)
CESCR	Committee on Economic, Social, and Cultural Rights
CPR	common-pool resource
CSD	Commission on Sustainable Development (UN)
DAC	Development Assistance Committee
DfID	Department for International Development (UK)
DKI	*Jakarta Daerah Khusus Ibukota* (Special Capital District of Jakarta)
DWAF	Department of Water Affairs and Forestry (South Africa)
ECOSOC	Economic and Social Council (UN)
ESP	Environmental Services Program (USAID)
GATS	General Agreement on Trade in Services
IBRD	International Bank for Reconstruction and Development
ICSID	International Centre for Settlement of Investment Disputes
IDA	International Development Association (or Assistance)
IDB	Inter-American Development Bank
IFC	International Finance Corporation
IFI	international financial institution
IMF	International Monetary Fund
IWRM	Integrated Water Resources Management
JBIC	Japan Bank for International Cooperation
KIP	Kampung (or Kampong) Improvement Program
LDC	less developed country
MDB	multilateral development bank
NRC	National Research Council (U.S.)
ODA	Overseas (or official) development assistance

OECD	Organization for Economic Cooperation and Development
OHCHR	Office of UN High Commissioner for Human Rights
PDAM	public water utility (Indonesia)
PPCP	public-private-community partnership
PPP	public-private partnership
PSD	private sector development
PSIRU	Public Services International Research Unit
PSP	private sector participation
PUP	public-public partnership
SEMAPA	water supply utility (Cochabamba)
SRI	social responsibility investment
TNI	Transnational Institute
UFW	unaccounted-for water
UNCITRAL	UN Commission on International Trade Law
(UN)CSD	UN Commission on Sustainable Development
UNDP	UN Development Programme
UNHCR	UN High Commissioner for Refugees
UNICEF	UN Children's Fund
UNSGAB	UN Secretary-General's Advisory Board on Water and Sanitation
UNWWAP	UN World Water Assessment Program
USAID	U.S. Agency for International Development
WB	World Bank
WCD	World Commission on Dams
WDM	World Development Movement
WHO	World Health Organization
WRI	World Resources Institute
WTO	World Trade Organization
WWC	World Water Council

DEFINING "PRIVATIZATION": A NOTE ON TERMINOLOGY

The correct terms to use when discussing water privatization are a matter of dispute. Terminology signals allegiances and thus is rarely neutral. The dispute over definitions reflects the slippery analytical terrain of water privatization debates and the inadequacy of conventional terminology to convey the complexities of urban water. In subsequent chapters, these ideas are developed in greater detail. Here, I simply clarify my use of terms.

How should we define *privatization*? Some (more frequently the proponents of private involvement) use a relatively constrained definition, reserving the term "privatization" for the sale of assets to the private sector—in other words, the private ownership of water-related infrastructure. In this case, the terms "private sector participation" and "public-private partnerships" are used to refer to a range of contracts whereby private companies build, manage, and/or operate infrastructure on behalf of governments. These contracts include concessions, management and service contracts, consulting services, and public-private partnerships with NGOs.

Others (usually opponents) use the word "privatization" as an umbrella term, to include the entire range of activities just mentioned. Although imprecise, this has the advantage of ease of reference. It also has the advantage of emphasizing the continuity between different types of private sector involvement, which all involve both the redistribution of governance to nonstate actors and the application of market-based norms, values, and practices in management and regulation. With a more general audience in mind, I use the term in this latter, inclusive sense; although I do not use the term to refer to broader trends of commercialization of water resources and services. But readers should bear in mind that, in most cases, the involvement of private sector companies in the infrastructure for drinking-water supply in urban areas—the focus of this book—involves what is technically termed "private sector participation" (and not full privatization).

The term "private" is also sometimes expanded to include the broad range of private entrepreneurs who run small-scale water businesses—a mainstay of water supply access in most cities around the world. These include, for example, water vendors in slums and developers who build stand-alone microtreatment systems for private, typically high-end residential developments. Here, I prefer the term "small-scale private entrepreneurs."

Some also use the word *private* to refer to community groups, religious associations, cooperatives, and nongovernmental organizations that are also extremely active in water supply, particularly to the poor. I think that this is unhelpful, as it extends the term "private" to cover all non-state actors. It is equally unhelpful, in my opinion, to characterize these groups as *public* in such a manner that we conflate their activities with that of the state. These groups (almost always not-for-profit) are more accurately characterized as community or non-governmental organizations of various sorts, and it is these terms that I have used throughout the book.

INTRODUCTION

PRIVATIZATION AND THE

URBAN WATER CRISIS

The Hague: March 2000

The World Water Forum—a global gathering held every three years to debate the world's most urgent water issues—is intended to be a solemn affair. But protests invariably disrupt the proceedings. The meeting in the Dutch city of The Hague was no exception: as Egypt's minister of Public Works and Water Resources began his inaugural speech, two audience members— one male, one female—suddenly appeared on stage. In full view of the gathered dignitaries and government ministers, the protesters approached the presidential table, removed their clothes, and handcuffed themselves together. Strategically scrawled on their bodies were the words "No to Water Privatization" and "Yes to Water as a Human Right."

Meanwhile, protesters in the audience (discreetly chained to their seats) shouted slogans accusing governments of colluding with private water companies to profit from the world's water resources. Some of their concerns related to the support given by conference organizers to private water companies and to their links with development organizations in favor of water privatization, including the World Bank. But protestors' slogans also targeted governments accused of environmentally destructive and socially inequitable water management.

The security guards were quickly overwhelmed, and the meeting ground to a halt. The protesters' message, captured by bemused journalists, was clear: water privatization had to be stopped, and government management of the world's water had to be dramatically reformed. But the Ministerial Declaration issued a few days later ignored these demands: the world's governments voiced support for private water management, making no mention of the human right to water or of the protesters' demands for environmental and social justice.

The events in The Hague are an example of the issues at stake in debates over water privatization and the world's urban water crisis. These debates have been heated, in part, because of the dramatic increase over the past two decades in private sector control and management of urban water-supply systems. During the 1990s, some of the world's largest multinationals (Bechtel, Enron, Vivendi) began expanding operation and ownership of water supply systems on a global scale: the largest private water company now has over 100 million customers worldwide. Although the vast majority of water supply systems around the world were (and still remain) government owned and operated, private sector activity increased at rapid rates. At the same time, many governments embarked on a series of market-based water sector reforms: the best solution, some argue, to the world's looming water crisis. This ethos has become increasingly widespread: water markets (and associated private water rights), private sector management, and commercial principles and practices have been introduced in the water sector worldwide over the past two decades.[1]

This has generated fierce controversy. Proponents of privatization assert that private companies will perform better: they will be more efficient, provide more finance, and mobilize higher-quality expertise than their government counterparts. Supporters also often argue that private involvement will facilitate broader reforms—such as the treatment of water as an economic good—that are required in order to ensure environmentally friendly outcomes such as water conservation and the reduction of pollution. These arguments rest on the claim that government management of urban water supply is beset by several interrelated problems: low coverage rates, low rates of cost recovery, low tariffs, underinvestment, deteriorating infrastructure, overstaffing, inefficient management, and unresponsiveness to the needs of the poor. This hotly disputed litany of government woes has dominated the discourse used by advocates of water supply privatization over the past decade, and is often summed up by the label "government failure."[2] From this perspective, it is unethical *not* to involve private companies if they can perform better than governments at providing water, particularly to the poor.

In contrast, opponents of privatization argue that government-run water supply systems, when properly supported and resourced, are more effective, equitable, and responsive; have access to cheaper forms of finance (and thus lower tariffs); and perform just as well as their private sector counterparts.[3] Those who reject privatization also warn of the negative effects—both social and environmental—of private ownership and management of water resources and water supply systems. Often they argue that it is unethical to profit from water, a substance essential for life and human dignity. David Harvey, for example, characterizes privatization of water supply as one example of "accumulation by dispossession"—the enclosure of public assets by private interests for profit—which invariably deepens social and environmental inequities.[4] Some go further, arguing that environmental protection and water conservation should be fostered through an ethic of water use, whether based on solidarity, scientifically determined limits to water use, traditional (often indigenous) water-use practices, or various forms of ecospirituality.[5]

This has obvious parallels with debates over public services—from health care to housing—and over the management of natural resources—from forests and fisheries to the global climate. From this perspective, water privatization is one example of a series of neoliberal-type reforms that have reworked the roles of welfare governments (and, in the global South, developmental states) in the provision of public services.[6] But water privatization inspires particularly fierce protest and, by the end of the 1990s, had become one of the most controversial issues debated in international-development and environmental-management circles.[7] Why would this be the case? One reason is that water fulfills multiple functions and is imbued with many meanings. Water is simultaneously an economic input, an aesthetic reference, a religious symbol, a public service, a private good, a cornerstone of public health, and a biophysical necessity for humans and ecosystems alike. It should thus come as no surprise that protests against water privatization have united a strikingly diverse range of movements: unions, environmentalists, women's groups, fair-trade networks, alternative-technology advocates, religious organizations, indigenous communities, human-rights organizations, antipoverty and antiglobalization activists. United in politically suggestive coalitions, these groups protest both privatization and the market-led water-governance reforms with which it is associated.

Another reason for the fierceness of protests is the fact that water is, in some sense, a final frontier for capitalism. Essential for life and (at least in the case of drinking water) nonsubstitutable, water throws up challenging barriers—technical, ethical, and political—to private ownership and management. The water privatization debate is thus a microcosm of contemporary struggles over the roles of states and markets, and over the acceptability and efficacy of markets and private ownership as mechanisms for public services delivery and as solutions to the world's putative environmental crisis.[8]

Urban Water: A Global Crisis?

Why, then, does this book focus on *urban* water supply? The answer is simple: the vast majority of formal private sector activity in water supply has taken place in urban areas, and the central promise made in the 1990s by proponents was that private companies would solve the world's urban water supply "crisis." The main features of this putative crisis are well known. Between 1950 and 1985 the proportion of the world's population living in urban areas doubled. But as cities have grown, urban services have not kept pace. The most recent estimate suggests that 970 million urban dwellers are without access to "adequate" water supply.[9] And the number of people without access to safe water continues to grow as rapid rates of urbanization continue in many parts of the world. The world's water crisis is thus, at least in part, an urban issue.[10]

Rural areas and outlying (or "peri-urban") settlements attracted little interest from private companies, as their small scale and low densities reduced profitability potential. Large urban centers were the focus of attention; since the prospect of profitability generally increases with the size of the urban area (because of important economies of scale), the urban bias of private sector participation is unsurprising. By the late 1990s, many capital cities of developing countries had committed to "private sector participation" contracts[11]—from Buenos Aires to Jakarta, Manila to Casablanca. The "global opportunity" offered to water companies by the world's water crisis (at least as depicted in industry rhetoric) was, in other words, largely to be found in cities: in managing, rebuilding, extending, and supplying urban water supply networks.

The bias of private investors and companies toward urban areas is borne out by the evidence. Of the world's total population, estimates suggest that only 3 percent are supplied via private operators, although this figure is much higher in some countries.[12] But when we look at cities, and particularly large cities, the picture changes: perhaps 20 percent of the world's urban population are supplied by the private sector, amounting to hundreds of millions of customers, most of whom became clients of private companies in the past two decades.[13] Urban water supply—the focus of the specific examples provided in this book—is thus the primary battleground over which water supply privatization is fought.

DEBATING PRIVATIZATION

Debates over privatization conventionally pit partisans of classic forms of government intervention against "neoliberals," whose reformulation of the role of the government emphasizes the need for selective regulation by the state, rather than direct state provision of public services.[14] Much of the debate between opponents and proponents of water supply privatization hinges on differing views about the role and extent of state versus market activity (or the "public" versus the "private" sphere). A range of political-economic arguments thus typically dominates water privatization debates, as with debates over privatization more generally.

But debates over water privatization also have an environmental dimension. Indeed, the arguments of water privatization proponents are perhaps best captured by the term "free-market environmentalism," a mode of resource regulation that offers hope of a virtuous fusion of economic growth, efficiency, and environmental conservation.[15] Supporters of "free-market environmentalism" argue that environmental goods will be more efficiently allocated and environmental degradation reduced or eliminated through establishing private-property rights, employing markets as allocation mechanisms, and incorporating environmental externalities through pricing. In short, markets will be deployed as the solution to (rather than being the cause of) environmental problems. Water services are sold on a commercial basis to customers, rather than supplied on an often-subsidized basis to citizens. Accordingly, the calculus of profit maximization—which at times leads to the prioritization of efficiency over other goals—becomes central to water governance.

In response, opponents of water supply privatization often frame free-market environmentalism as a form of "green imperialism" or "green neoliberalism." They point to studies that have demonstrated the limits, unexpected consequences, and impacts of neoliberalizing nature in a broad range of historical and geographical contexts.[16] They argue that while environmental degradation (an inevitable by-product of capitalism) may be mobilized as an opportunity for continued profit,[17] the involvement of private companies will not necessarily ensure an overall improvement in environmental quality; on the contrary, companies are likely to engage in cost-cutting measures detrimental to environmental health, dignity, and well-being.

As explored in later chapters, views from developing countries offer different perspectives. In rural areas, often rooted in indigenous water-use practices, communities offer cultural as well as political-economic critiques of *both* private and government provision of water. These critiques are echoed in urban slums and unserviced city outskirts ("peri-urban" areas), where communities and small-scale, unregulated private businesses (rather than government or large-scale private companies) play the lead role in supplying water on a daily basis. The public-versus-private debate plays out, here, on a complex terrain.

REFRAMING PRIVATIZATION

The preceding discussion implies that we need to reframe the question of privatization in two ways: to examine privatization as an environmental as well as a socioeconomic phenomenon; and to integrate an analysis of privatization with an understanding of the simultaneous and often overlapping roles played by government, private companies, and community actors. This is the central analytical task of part 1 of the book.

But before I continue, there is a caveat. Although much of the literature begins from a strongly pro- or antiprivatization stance, this book begins from a different starting point. I argue that conventional models of both government and private provision have serious flaws: as I mentioned in the preface, examples of well-run public and private water supply systems, as well as examples (sadly, more abundant) of poorly run public and private water supply systems, can be found around the world.[18] Accordingly, we cannot categorically refute private sector involvement in water supply, nor simplistically defend government provision. Rather, I suggest that we need to expand our focus beyond formal water supply networks, develop an understanding of the roles that both public and private actors play in governance of urban water supply for the poor, and pay closer attention to the practices of urban water use in developing countries (particularly those of the "urban unconnected"). This requires rethinking some of the concepts on which the water supply debate is conventionally predicated, particularly the terms "public," "private," and "community."

Why is it helpful to rethink these terms? In subsequent chapters, I argue that the debate over privatization is not well served by concepts derived from what Charles Taylor terms our "modern social imaginary," which assumes a clear division between a public (governmental) and private sphere, adjudicated by mechanisms of popular sovereignty.[19] In successive chapters, I will provide examples of why conventional concepts of public and private are inadequate for describing the complex interrelationships between communities and water use. Further, I argue that the debate is predicated on a concept of popular sovereignty that fails to account for the ways in which many communities actually manage water access.[20]

My analysis also emphasizes the environmental dimension of water supply privatization—an issue often glossed over in contemporary debates. Environmental issues are both a driver and a source of critique of privatization: for example, the poor quality of drinking water serves as a justification for privatization, and concern over the impacts of privatization on fresh water is often a central concern of privatization opponents.

Urban water supply is, in other words, an environmental issue as well as a social and economic one. This is rarely recognized in the literature on privatization, or indeed on urbanization. And where urban nature does receive attention, it tends to be framed in terms of environmental conflicts or green enclaves (such as parks). In contrast, I approach urbanization as simultaneously natural and social: constituted by (and constitutive of) political ecological processes. This implies a rather unorthodox view of urbanization (and here I rely heavily on the work of David Harvey, Roger Keil, and Erik Swyngedouw), which does not circumscribe urban nature to "green spaces," but rather focuses on the material flows—such as excreta, water, wastes—that move through the city, and the different governance processes, power relations, infrastructures, and subjectivities via which these are mediated.[21]

Urban nature is, in other words, visceral, embodied, and woven through the fabric of the city. This occurs, of course, in highly differentiated patterns: the urban poor, who live within the interstices of the city (in floodplains and along riverbanks, on steeply eroded slopes and marshy land) often experience water as a threat to physical safety, both in terms of flooding and poor water quality. Better water governance—addressing ecological concerns across urban watersheds—would tackle these issues. Environment-related water concerns are thus an imperative, and not a luxury, for the urban poor; but much of the debate about water supply privatization has not adequately captured these broader concerns.

These arguments are developed throughout the book, as I attempt to rethink the terms "public," "private," and "community," and explore how these terms might be inflected with an ecological sensibility to refine our understanding of the contributions and limits of communities, states, and markets (as conventionally understood) in achieving social and ecological justice. The arguments that flow from this perspective, although focused on drinking water in urban areas in developing countries, are intended to speak to broader debates over the respective roles of states, markets, and communities in economic life; our collective response to environmental crises; and the role of civil society (or the "public sphere") in adjudicating questions of social and ecological justice.

The focus of my arguments is thus on conceptual rather than practical aspects of the water privatization debate. Of course, the broad range of practical issues (such as labor and environmental standards, tariffs, financing, regulation, technologies, and social policies) is important. But these issues are already well addressed in the literature, and I make reference to key sources as appropriate. My task, rather, is conceptual: to sketch out the ideological arena in which the debate over water supply privatization takes place, to provide historical context for the emergence of privatization, and to reframe our understanding of the involvement of governments, communities, and private actors in water supply as a means of suggesting new ways of thinking about—and eventually formulating solutions for—the world's urban water crisis. The concept of "governance failure," which I introduce in part 1, is central to this latter task.

Let me summarize my approach by outlining three questions that successively structure my analysis. First, why has privatization emerged as an increasingly widespread mode of water supply management, and what are the arguments made by its proponents and opponents? Second, can privatization fulfill proponents' expectations—particularly with respect to water supply for the urban poor in developing countries? Third, given the limits to privatization, what are the alternatives? How, in other words, might we move "beyond privatization" both conceptually and practically? The first two questions are the focus of part 1, and the third question is dealt with in part 2 of the book.

PART 1: DEVELOPMENT, URBANIZATION, AND THE GOVERNANCE OF THIRST

The polarization of the water privatization debate rests on widespread yet often implicit assumptions about the failings of governments and markets. Proponents of the "market failure" argument, for example, argue that private companies' drive for profit necessarily compromises their management capacities. Proponents of the "state failure" argument, in contrast, tend to argue that governments are inevitably unaccountable and unresponsive to the demands of citizens for public services. My analysis begins from a different starting point, through focusing on issues of governance, many of which are common to both government and private companies. Governance, here, is defined as a practice of coordination and decision making between different actors, which is invariably inflected with political culture and power. This definition of governance is not the norm in the literature on water management, which tends to constrain the definition of governance to a narrowly technical decision-making process. Defined in this broad way—as an expression of social power—I argue that the concept of governance can help us understand some of the persistent failures of government and private models, and the emergence and persistence of fragmented patterns of urban water supply.

Reframing the Urban Water-Supply Crisis

Chapter 1 explores the multiple modes of water supply access in contemporary cities, and cautions against focusing solely on formal networks, which reach only a small portion of urban residents. To frame this discussion, I present three urban water supply models—government, private, and community—each associated with a range of different technologies and delivery methods (from on-foot water vendors to bottled water, private wells to public standpipes). Each of these three models has its flaws, which I explore through the concepts of governance failure, state failure, and market failure. The latter two concepts are widely discussed in the water privatization literature: proponents of privatization generally articulate "state failure" arguments, and proponents of public water usually espouse "market failure" arguments. Here, I introduce a third concept, "governance failure." Subsequent chapters develop the interplay between models of water supply (government, private, and community) and these three "failures" (state, market, and governance) as a means of explaining the roots of the urban water supply crisis, and thereby the origins of water supply privatization and its impacts.

Constructing "Public" Water

Chapter 2 provides historical context for the analytical arguments presented in chapter 1. Supplying water to urban areas has been high on the agenda of the international community since the United Nations Water and Sanitation Decade (1981–90), during which bilateral aid and multilateral finance were directed toward water supply projects in unprecedented amounts.[22] At the end of the decade, more people (in absolute terms) enjoyed "improved water supplies" than ever before, yet in many countries supply failed to keep pace with population growth and with accelerating rates of urbanization over the latter half of the twentieth century.[23] Today, the scale of the problem is seemingly beyond our best efforts: even the ambitious Millennium Development Goals call for a reduction of only 50 percent in the number of those without sustainable access to safe drinking water.

In order to explain how this situation arose, chapter 2 explores the evolution of water-related development policies in the post–World War II period. In the 1950s and 1960s, development agencies advanced a specific vision of water management: a "modern integrated ideal" predicated on large-scale hydraulic works, particularly large dams. Water-supply delivery systems were marginal to this approach, and gained attention only from the 1970s onward. When water supply did become the object of significant lending efforts, development finance tended to focus on urban areas, and a "municipal hydraulic" model emerged: large-scale water production facilities, linked to integrated water networks targeted at economically productive, high-value urban neighborhoods, governed by centralized water-supply-services providers—either municipal departments or (more rarely) public corporations.

The urban water-supply crisis is in part the result, I argue, of the ways in which this "municipal hydraulic" paradigm has been applied (and mishandled). Chapter 2 analyzes the involvement of the World Bank in lending to government-run water projects as a means of illustrating this point. This assessment explains, in part, the drivers for water sector policy reforms within the World Bank's water policy, which set the stage for the privatization agenda. The Bank's experience is, I argue, emblematic of broader trends as a whole; by the end of the century, many bilateral aid donors and international financial agencies were engaging in a range of activities designed to facilitate—or even impose via "conditionalities"—private sector management of urban water-supply systems.[24]

Debating Private Sector Involvement in Urban Water Supply

Chapter 3 traces the expansion of private sector activity in the water sector in the late twentieth century. It, too, begins with a historical discussion and demonstrates that although government management has dominated the water sector, private companies have long been active in water supply—although their legitimacy and impacts have been hotly contested. The chapter explores why private sector activity increased so rapidly in developing countries in the 1990s. Throughout this period, private companies (usually multinational water-services firms based in a few developed countries, notably France and the United Kingdom) sought to sign long-term contracts with (usually municipal) governments for the maintenance and extension of urban water supply networks. But some private companies—most importantly the large multinationals—have recently strategically retreated from certain regions of the world and have significantly reoriented their investment and growth strategies. The chapter summarizes recent evidence that suggests that the mainstream "private sector participation" (PSP) model—of large water multinationals engaged in long-term concession contracts—is likely to be more limited in scope than previously thought. Indeed, over the past few years, an increasing consensus has developed that private sector participation in water supply will not be able, as some proponents had hoped, to succeed where governments had failed in providing "water for all." The chapter suggests that private companies encountered many of the same barriers—and displayed many of the same management patterns—as their public counterparts, and concludes by exploring the debates that this has engendered.

Citizens Without a City

In chapter 4, I illustrate the arguments made in previous chapters through a case study of Jakarta, Indonesia. I explore how urban water management entails the "production of thirst," and document the ways in which cultural norms, political commitments, and the seemingly mundane practices of water managers combine to exclude the poor from accessing water. The analysis emphasizes the political rationalities that underpin the creation of differentiated infrastructure systems. Simply put, access to services such as water is the subject of political negotiation, mediated through identity, urban infrastructure, and the differentiation of urban space. The involvement of the public water-supply company in creating a highly fragmented water-supply system, and the troubled track record of the private companies invited into Jakarta in the late 1990s, illustrate the point that both public and private companies encounter significant "governance failures" in attempting to extend urban water supply to low-income households and neighborhoods.

The case of Jakarta is not unique: fragmentation of access to water supply infrastructure is common in cities around the world. I argue that these failures to provide universal water supply are not mere lapses, but are systemic and structural (although not always deliberate or planned) outcomes. In those cities, as in Jakarta, absolute water scarcity is rarely (if ever) the reason why people in cities are without access to safe water. My intent here is not to deny the reality of the world's urban water crisis, but rather to question its causes, and to encourage skepticism about the rhetorical uses to which the specter of "crisis" can be put (particularly when it serves as a justification for privatization).

PART 2: BEYOND PRIVATIZATION: DEBATING ALTERNATIVES

Part 2 turns to the debate over alternatives to privatization, and to our conventional models of public services provision. The partial retreat of private companies from the business of water supply in the past five years has intensified debate over the appropriate role of the private sector in supplying water. Even ardent proponents of privatization admit that the private sector will not supply "water for all." Meanwhile, opponents of privatization have taken an increasingly critical look at conventional models of government provision and regulation of water. Conceptually, the current period is thus ripe with possibility, as the debate is more nuanced. Politically, as with more general debates over postneoliberalism, the situation is subtler but also more confused, with few truly progressive alternatives in sight. The financial crisis of deepening severity had, by late 2008, lent greater urgency to this situation.

Practically, this has led to a resurgence of interest in alternatives to water privatization, which is the subject of part 2. Successive chapters critically examine the most politically popular alternatives in contemporary debate. Chapter 5 discusses the transnational struggle over the human right to water and points to some of the potential limits of this approach, while suggesting that other notions of rights (such as the "right to the city"), predicated on political struggle, will provide more traction in obtaining "water for all." Chapter 6 interrogates notions of commons and community water supply often proposed as alternatives to both public and private provision. Chapter 7 speaks to ecological issues and the (often overlooked) environmental dimensions of water privatization.

Some of these alternatives represent only modest modifications of government or private models (such as small-scale private water companies, or so-called government–private community partnerships). Others, such as proposals for a human right to water or legal reform to create water commons, represent more radical change, and reorient the conceptual terrain and practical focus of debates over solutions to the global water crisis. It is because of their radical nature that I have chosen to focus on this (not necessarily compatible) triad of concepts: human rights, community, and ecological governance.

The Human Right to Water

Chapter 5 documents the emergence of a global campaign for a human right to drinking water and points to the significant conceptual and practical limitations of this concept. There exist, as the chapter explores, considerable technical difficulties related to the implementation of the human right to water. And there are also more profound criticisms to be made. First, a potentially irreconcilable tension arises between the human right to water and traditional (communal) water rights—which are particularly important in places with indigenous populations. The chapter also raises the issue of the anthropocentrism of human rights, which excludes ecological rights (the rights of nonhumans). Providing a human right to water may, ironically, enable justifications for the further degradation of hydrological systems on which ecosystems (and, of course, human beings) depend. A third criticism is that the framework of human rights is individualistic and legalistic, and hence cannot address the collective governance issues that constrain access to water on the urban scale. The equitable provision of water supply necessarily implies a degree of solidarity (both physical and material). Yet I argue that it is precisely this notion of solidarity that human rights, in isolation, cannot provide.

Nonetheless, the chapter also recognizes that the human right to water is a necessary and useful (although perhaps not sufficient) strategy for solving the world's urban water crisis. The importance of the aspirational dimension of a human right to water should not be underestimated: it raises expectations and places responsibility for those expectations on both public and private actors. As a simple example, the burden of government regulatory oversight of private actors would likely be higher in a context where the human right to water was legally recognized. Moreover, a human right to water shifts the focus from the putative advantages of public versus private providers to the responsibilities and accountability of all actors involved in water supply—both network and nonnetwork providers. The flexibility implied in this approach does not, however, mean lower standards; on the contrary, a human right to water demands certain minimum levels of services, and thresholds for availability, quality, accessibility, and affordability. As a result, a human-rights approach implies a focus on the most vulnerable groups, and thereby provides a potentially powerful means of combating the "elite capture" of water supply systems. A "rights-based" approach to water thus has both potential strengths and potential pitfalls—as explored in the case study of South Africa provided in chapter 5.

The Ambiguous Merits of Community Water Management

Chapter 6 turns to the question of community water supply. For anti-privatization advocates, community water supply is a strategy of placing water in the public sphere as a means of opposing privatization. For proponents of the private sector, involving a community is a means of resisting government interventions. Both are often equally suspicious of government action and cognizant of the benefits of involving communities in the management of water.

Of course, appeals to community are not unique to the water sector (and flourish under a variety of banners, from postneoliberalism to the "post–Washington Consensus"). Indeed, alternative community economies of water exist around the world and have been the focus of extensive study. Chapter 6 does not review this literature; instead, the analysis focuses on a critique of the concept of community invoked in current debates over water privatization. The chapter summarizes the rather eclectic range of community options proposed in current debates, from water "commons" and cooperatives to various forms of community water governance. In the analysis, I suggest that although community provision has many advantages, it cannot comprehensively deal with the demands of distributive justice nor adequately address environmental concerns in urban areas: hence the "ambiguity," in my opinion, of appeals to community. A brief case study of Bolivia illustrates these points.

The Environmental Dimensions of Privatization

The thrust of chapter 7 is that debates about privatization need to systematically integrate environmental as well as socioeconomic concerns. For the urban poor, issues of livelihoods and environmental quality are intertwined. But the focus of much of the privatization debate obscures the larger environmental picture, in which improvements in the quality of human life are predicated on the restoration and preservation of water ecosystems. From this perspective, ecological efficiency, the conservation of natural resources such as water, and the eradication of poverty are necessarily interlocking goals.

These observations are inspired by debates over what academics term "political ecology," a mode of analysis which focuses on the interrelationships between livelihoods, justice (both environmental and socioeconomic), political economy, and sustainability concerns. Such views, and others like them, begin from the point of view that resources are cultural and ecological (and not just technical) phenomena. Simply put, just as we admit that forests are more than wood warehouses, we must recognize that waterways are more than natural reservoirs. The implication follows that innovations in our management of resources—such as privatization—have simultaneously socioeconomic, cultural, and ecological impacts that must be weighed together. In chapter 7, I use the term "ecological governance" to capture these ideas and expand on their consequences for debates over water privatization, properly framed as an issue of socioecological justice.

Reframing Privatization

Why do these arguments matter? As explored earlier, the water privatization debate intersects with broader debates over development, the roles of states and markets, and environmental management. Disagreements over the support given to private companies by development agencies raise questions about the role and extent of markets and private companies in the provision of public services. Debates over the involvement of private companies raise issues of social, economic, and ecological sustainability. Questions about the respective roles to be played by communities, governments, and private corporations raise issues of deliberative democracy (which might be simply defined as theories of democracy that treat collective, public deliberation of ideas as the core of the practice of democracy, rather than the mere aggregation of private votes). Essential for life, fresh water provides a powerful lens with which to examine these broader debates on the legitimate roles of governments, markets, and communities in environmental management and the provision of public services.

These conceptual issues are not restricted, of course, to urban water supply. Similar debates are occurring with respect to public services (from education to health care) and utility networks (from electricity to information technology). But water is a particularly interesting issue because it is essential for life and ecological health and is imbued with spiritual and political significance. Water supply thus raises issues of social and ecological justice in a particularly acute way. In debating water privatization, in other words, we are also debating the relations and responsibilities among private actors, communities, governments, and the environment.

Opinions adopted within these debates will naturally be the subject of fierce dispute. I do not expect that all readers will be convinced by my analysis (and indeed I expect that disagreement is inevitable, particularly with my critiques of the private sector in chapter 3, the human right to water in chapter 5, and community water supply in chapter 6). Readers, however, may like to bear in mind that my objective is not to convince them of the definitive failure or success of water privatization initiatives, nor to advocate for any one particular alternative. Rather, the book's primary goal (and a measure of its success) is to draw the reader into a broader conceptual space in which our criteria for alternatives to conventional public and private approaches (which I believe to be urgently necessary) can be refined, and in which wide-ranging debate over urban water governance can flourish.

These are urgent matters. Over half of the world's population lives in cities.[25] In lower-income countries, many urban residents lack access to safe water. Most poor households lack in-home connections to the water supply network. Instead, the urban poor rely on alternative solutions—like water vendors or self-dug wells. These alternatives are often costly and provide water of poor quality and unreliable availability. The public-health implications—high rates of child morbidity and mortality—are staggering. The impacts on people's productivity, longevity, and dignity are profound. Supporters of privatization argue that private companies may succeed where governments have failed in supplying water to the urban poor. Interrogating this argument, demonstrating its limits, and exploring alternatives are the central tasks of this book.[26]

DEVELOPMENT, URBANIZATION, AND THE GOVERNANCE OF THIRST

1

GOVERNANCE FAILURE

REFRAMING THE URBAN WATER

SUPPLY CRISIS

Consider the following story of three neighbors on the outskirts of one of the world's megacities. It suggests that concepts like "public" and "private" are of little use in describing the reality of the daily struggle for water supply in much of the world. It also suggests that communities play an important role in the complicated web of relationships that mediate water supply access in cities.

Ani lives in a gated community, with a direct road connection to the highway that runs into the center of the city. Like her neighbors, she used to be connected to the municipal water supply system. But then she became frustrated by frequent cut-offs, low pressure, and the sometimes murky, smelly water. Several of her neighbors bought expensive booster pumps, but these helped only a little. So last year, her neighbors decided to create their own water cooperative and they drilled a series of deep wells. This was expensive, but the water is free and so plentiful that Ani has recently installed an imported whirlpool bathtub and plans on building a swimming pool in her backyard. Bottled water from a local company that taps several springs in the hills above the city (much to the distress of local farmers, who are thus deprived of their customary water sources) satisfies the family's daily drinking water needs. For guests, and as a special treat, Ani buys mineral water imported from Europe.

Mira lives in a small house along the busy main road that passes by the gates of Ani's neighborhood. Her family has converted the front of their home into a shop that sells snacks, cigarettes, and bottled drinks to passing drivers. She considers herself lucky to have a connection to the municipal water supply network, which few of her neighbors have. But she has experienced the same water problems as Ani. So although Mira uses tap water for washing dishes and doing laundry, every few days she buys bottled water for drinking. This requires strapping twenty-liter water jugs onto the back of the family scooter and driving to the private microtreatment plant that opened a couple of years ago down the road, where she waits in line for her turn at refills. Mira knows that the plant is unregulated, and she worries about the quality of the bottled water she buys, but since her husband lost his job during the last economic crisis, "local" bottled water is all her family can afford.

For additional income, Mira regularly fills water bottles from her kitchen tap and sells them to her poorer neighbors. Water runs for only a few hours a day, so the tap is left constantly open; when the water starts flowing, she rushes to fill buckets and other containers as quickly as possible, day or night. Mira charges her neighbors double the rate she pays, which is still much cheaper than "factory-made" bottled water. It is usually the local women who collect the water, some bringing empty plastic bottles they have scavenged from the nearby dump.

Alia is one of Mira's regular customers. With her family, Alia lives in a small, self-built shack, illegally occupying the thin strip of land between Mira's house and the river, one of the main sources for the city's water supply network. As the city's sewerage system does not extend to the neighborhood, and few houses have septic systems, the river functions as an open sewer. The most impoverished households scavenge wood scraps to construct "helicopter toilets" over the water to provide themselves with a bare minimum of privacy and comfort. Downstream, the water is so polluted that local farmers have had to stop using it. And the neighborhood's groundwater is predictably polluted. Nonetheless, Alia uses her family's shallow hand-dug well for washing and bathing water, as she can not afford to buy bottled water for all of their household needs.

When the dry season comes, or when Mira has no more tap water to sell (which happens often), Alia pays a water vendor to deliver water to her house. The vendor fills twenty-liter jugs at the nearest functioning public tap, which is over a mile away. He then delivers the jugs in a wheelbarrow. The drier the weather, the higher the prices demanded by the local "chief" in charge of the ostensibly public tap, and the more expensive the water. At the height of the dry season, Alia pays up to ten times more per liter than Mira pays for her tap water. This year, Alia's husband installed a small cistern on the roof, in hopes that the rainwater they gather will help them reduce their water bills, which sometimes take up to one quarter of their family income. Many of Alia's neighbors have recently installed illegal connections to the nearby municipal water pipe (the cause of the low pressure in the network), but Alia refuses to do so, because she is worried about being evicted by government inspectors.

No one expects that the situation will change any time soon. Government promises made over the years have not been kept. The recent transfer *of the management of the municipal water supply system to a private water company, the source of much protest, has yet to have any impact. Ani has noticed that the local, name-brand bottled water that she buys has recently become more expensive, but doesn't know why. In fact, the water bottling company, under recent government reforms that created a series of regional water markets, has now started purchasing water "rights" from local farmers in the watersheds surrounding the city. Ironically, it and other private bottled water companies outbid the city's water supply department. Because it can't afford to access other, new water supplies (which are far away and much more expensive), the government has put plans to expand the city's water supply system on hold indefinitely.*

The story of the three big-city neighbors mirrors the daily reality of many urban residents around the world. Both rich and poor households, on any given day, use several different types of water: groundwater for bathing, bottled water for drinking, and rainwater for washing laundry. They interact with a range of different water providers: the municipal water supply network, the tanker-truck driver, the local water vendor, and the bottled water seller. And they use a variety of supply technologies, from highly industrialized to artisanal: individual household wells or rainwater collection systems, reticulated water networks, water pumps, gravity-fed roof tanks. Acquiring water is a complex and time-consuming task, and it requires intimate knowledge of the political ecology of the city's water: where it flows at different times of year, how much it costs and how those costs vary, how trustworthy different suppliers are, and how much water quality varies across time and space. This bears little resemblance to the "flush and forget" experience that many wealthy urban residents take for granted, predicated on a uniform, standardized, universal model of networked water supply provision that citizens of lower- and middle-income countries are presumed to aspire to.

As this suggests, the terms "informal" and "unserviced" are unhelpfully imprecise. Similarly, the term "network," and the interconnectedness it evokes, is a poor descriptor of water supply systems in many cities. Government networks, if they exist, are limited in extent. Government and private networks coexist and compete with informal types of provision. So definitions of "public" and "private" are more complicated than we often realize, and this complexity is associated with the diverse types of technologies used for supplying water.

How can we explain this fragmentation, at the level of both the network (which weaves its way unevenly through communities) and the individual household (which relies on multiple types of water supplies)? What are its causes and its impacts? What does it suggest about the experience of modernity and urbanization? These questions are of direct relevance to the water privatization debate because both the private and public sectors are frequently blamed for the failure to achieve an integrated network. In subsequent chapters, I will return to a case study of patterns of fragmentation and the different processes of exclusion that create them. Meanwhile, in this chapter I want to set the stage for the analysis through an exploration of the multiple ways in which we govern urban water.

What follows below is a simple framework that describes the governance of urban water, on the one hand, and the technological approaches used to supply water, on the other. I identify three models of water management that exist in contemporary cities—private, state, and community-run—and explore how they overlap and interact with one another. As a means of introducing a conceptual framework for analyzing exclusion from access to public services, I discuss these models with respect to concepts of market failure, state failure, and governance failure. I argue that all of these failures—market, state, and governance—affect the provision of urban water supply and play a role in creating the contemporary urban water crisis. In subsequent chapters, I elaborate on these points, and trace the emergence and outcomes of these models; here, my discussion is more abstract and focuses on the mechanisms and pathways by which exclusion from urban services occurs, mediated by governments, private actors, and communities.

Urban Waterscapes: Networks versus Archipelagos

Understanding the diversity of types of urban water provision begins with the recognition that multiple modes of water supply coexist in many cities. For example, private and government provision coexist. Government networks circulate through the city, but private companies supply bottled water, and private water vendors deliver water (often taken from the public network) to people's homes. Similarly, industrial and artisanal modes of production coexist in many cities. Many households will use, for example, rainwater, private wells, and networked water supply, sometimes simultaneously.

In many cities, then, water supply networks do not operate homogeneously over the urban landscape. Rather, they overlap with what policymakers term "alternative service delivery mechanisms" or "small-scale independent providers." In this sense, the metaphor of the archipelago—spatially separated but linked "islands" of networked supply in the urban fabric—is more accurate than the term "network." The distribution of this archipelago, as we shall see in subsequent chapters, is highly correlated with household income. Only a minority of urban residents in developing countries will have access to a networked water-supply connection within the home, and many of these residents will not rely exclusively on the network for all of their water supply needs. In nearly all cases, the local government typically fails to extend public services to socioeconomically marginal areas of the city. There, water pipes are thinly scattered and in many areas are completely absent. Poor neighborhoods are less likely to have water mains, and there is much less distribution (or "tertiary" pipes) even where primary mains are present. The sight of a large water main running through a poor community in which the adjacent homes do not have household or even neighborhood outlets is not unusual.

"Splintered" Urban Networks

Why is this fragmentation of urban water supply networks so persistent in cities in developing countries? This question presupposes a certain ideal: that of universal networks. When compared against this ideal, the provision of public services (water, electricity, telephones, etc.) in many cities is indeed fragmented. A casual observer might assume that this fragmentation is an aberrant and perhaps recent phenomenon. To cite an influential example, Stephen Graham and Simon Marvin argue in *Splintering Urbanism* that over the past two decades cities have suffered from a fragmentation of access, control, and pricing of network infrastructure, including water supply.[1] This "splintering," they assert, has occurred along with the restructuring of utility networks, embedded in wider changes in aid and financial flows, technological innovation, social attitudes, and governance—particularly the reconfiguration of citizens' entitlements in light of new understandings of the appropriate role of the state in services provision.

Graham and Marvin conclude that this fragmentation is due to political-economic processes—such as neoliberalism—that have changed the way cities are governed and services provided. Their analysis is representative of much conventional thinking about cities and public services, which holds that cities are splintering because of macroeconomic and political trends, such as high levels of debt, that reduce the fiscal capacity of the state to provide and extend government services. According to this view, these trends create the ideal conditions for "vulture capitalists" (in our case, private water companies from the North) to aggressively seek out new markets.

The counterargument, which dominated the discourse of mainstream development organizations in the 1990s, is the assertion that poor government management is the cause of fragmentation. Lack of incentives for efficient management, self-interested behavior on the part of politicians and workers ("rent seeking"), corruption, underinvestment, and a lack of necessary expertise are frequently offered as reasons for the poor performance of government-run water supply systems. The latter arguments have had the upper hand in international debates within the development community for the past two decades.

Networks as "Elite Archipelagos"

In both sets of arguments, the modernist ideal of large technical systems and integrated networks holds sway. Development interventions, whether in support of government or private involvement, are predicated on this notion, from which cities of the South are seen to deviate, but to which they are expected, eventually, to conform.

For those who actually work in the cities of the South, however, these arguments are both incomplete and inaccurate.[2] Like much work that engages with questions of the relationship between society, cities, and infrastructure, Graham and Marvin's work is rooted in a Northern context. Their concern with the social and environmental impacts of the "collapse of the integrated ideal" presumes the existence of modern infrastructural networks with widespread, if not universal, provision. Although they acknowledge the differentiation of service provision (which they characterize as "spatial apartheid"), their concern with the interrelationship between splintering infrastructure networks and fragmentation of both urban space and social consensus is based on a narrative which assumes the prevalence of the "modern networked city" as a generic phase or stage of urban development.

Urban theorists working in cities of the South have explored alternative frameworks for understanding "urban informality" (or, as Ananya Roy terms it, "medieval modernity").[3] From this perspective, urban informality is a mode of metropolitanization, of which fragmented access to urban services is a material expression in urban infrastructure. Scholars such as Roy, Arjun Appadurai, and Partha Chatterjee have explored the political dimensions of urbanization and exclusion. Common to these analyses is an understanding of the urban form of cities—and particularly "slums" (or "favelas" or "informal settlements")—in the global South as purposive, self-organized, economically productive, and articulated in a myriad of ways with "modern" parts of the city. This view contrasts sharply with depictions of urban slums (as in, for example, Mike Davis's much-criticized best seller, *Planet of Slums*) as being exclusively constituted of zones of exclusion: chaotic, disorganized, abandoned by the state.[4]

These critiques suggest that we should pay closer attention to urban informality and direct our attention to the broader range of community and corporate organizations involved in the provision of water supply. In doing so, the inadequacy of conventional terms (corporate, public, private) becomes apparent. For example, we might use the term "corporate" to refer not only to the private sector, but also to both government and private suppliers, organized in formalized management hierarchies, with associated forms of expertise suited to the management of large-scale technical systems (such as, for example, engineering expertise). Corporate providers can thus be state or private. Of course, there are important differences between them, as I shall explain. But they share many notable similarities in terms of technological expertise and business models.

This definition of "private" provision is blurred even further when we consider the informal, often unregulated, small-scale water supply businesses that meet the majority of people's water supply needs in many cities. Are these private? They often work alongside community organizations, which are both nongovernmental and not-for-profit. Both often operate with community goals in mind, with a self-identified ethos of public service. So they exist in an uneasy, interstitial space within the public-private debate. "Public" here does not refer to government but includes aspects of what we might term, again, "civil society." Of course, the unions representing water workers—whether within a government or a privately run company—often assert that they defend "public" water, in the interests of civil society.[5]

This framing of the problem of urban water supply suggests that we need to pay more attention to community strategies of service provision. I admit that there is nothing inherently new in this argument. The approach taken in this chapter has much in common with debates (in both academia and development policy) over institutions and social capital. Leaving aside obvious criticisms of these terms, we can simply note their common attempt to describe the residual lying beyond the dualities of governments and markets. This echoes tripartite characterizations of modern society offered by a variety of thinkers, from Arendt (private, public, and the social realm) to Hegel (family, civil society, and the government). These frameworks suggest that dichotomous models of modern society, when thought through seriously, tend to break down.[6] Table 1.1 presents a generic typology of this tripartite approach to the ways in which we govern resources and, more specifically, urban nature.

The preceding discussion also suggests that the distinction between corporate (both government and private sector) and community may have greater analytical utility than the "public" versus "private" binary. I shall

TABLE 1.1. *Governing Nature: Three Models*

	Government	Private corporation	Community
Primary goals	Safeguard public interest	Maximize profit	Serve community interest
	Conformity with legislation/policy	Efficient performance	Effective performance
Property rights	Public (state) property	Private property	Commons or private property
Organizational structure	Municipally owned utility or municipal department	Private company (corporation)	Community cooperative (association/network)
Regulatory framework	Command and control	Market mechanisms	Community-defined goals
Accountability mechanism	Hierarchy	Contract	Community norms/sanctions
Key incentives	Voter/taxpayer opinion	Price signals (share movements or bond ratings)	Community opinion
Key sanctions	Political process via elections	Customer opinion	Livelihood needs
	Litigation	Financial loss	Social pressure
		Takeover	
		Litigation	
Consumer role	User & citizen	User & customer	User & community member
Governance (decision-making process)	Expert-led, top-down	Individualistic, distributed	Participatory, bottom-up
Primary decision makers	Administrators, experts, public officials	Company managers, experts	Leaders and influential members of community
Worldview of nature	Public good	Commodity	Common-pool resource
	Source (economic input) and sink (waste disposal function)	Source (economic input) and sink (waste disposal function)	May have symbolic/religious dimension

return to this distinction between corporate and community control in subsequent chapters, as well as to the question of civil society in various guises.
For now, I will merely emphasize the fact that simple public-private dual-
isms break down when confronted with the complexity of water supply
provision in the cities of the South. The conventional model of "privatization" deployed since the 1990s, for example, usually is understood to
pertain only to the formal water-supply network. Figure 1.1 illustrates this
complex array of ways of accessing water.

The different modes of water supply access described in figure 1.1 suggest that, for the urban elite, water supply is often relatively abundant and
relatively cheap. But for the urban poor, the scarcity of potable water is a
daily hardship. Wealthy, mostly white South Africans, for example, use on
average six hundred liters per person per day of potable water delivered
through taps inside the home, whereas the poor and mostly black residents
of peri-urban settlements may average as little as ten liters per person per
day, often walking several kilometers to fetch water. This unequal pattern is
common throughout the developing world. But here it is important to note
that the "urban unconnected" are not limited to low-income households.
Many wealthier communities, particularly in suburban locations, are also
without network access. As the example at the start of this chapter suggests,
however, the capacity to access adequate water supply is highly variable
and often predicated on income. And the absence of an integrated network
eliminates, of course, the possibility of cross-subsidization through water
bills, which underpinned the universalization of water supply networks in
most wealthy countries.

Figure 1.1 also suggests that a range of private providers operate within
the urban landscape. In some cases, they employ sophisticated technologies
(such as water supply networks); in others, artisanal technologies (such as
shallow hand-dug wells and hand-carried water jugs). These latter, small-
scale private providers are the primary means by which lower-income households access urban water supply. In the absence of an integrated network,
individuals and communities respond with solutions of varying degrees of
equity and environmental sustainability.

GOVERNING WATER SUPPLY

The utility of the typology presented in figure 1.1 becomes evident when
we focus our attention on lower- and middle-income households in urban
areas, usually concentrated in slums or the "peri-urban" areas beyond the

1.1. Modes of urban water supply (adapted from Bakker 2003)

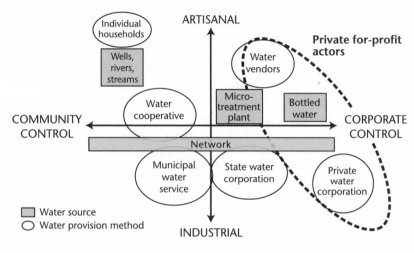

city center. Here, the distinction between public and private is often unclear. Property rights may be absent, informal, or customary. Governments do not provide (and indeed have never provided) universal public services. "Public goods"—from health care to water supply, from education to electricity— are circumscribed to the elite. These services are thus more accurately described as both material emblems and cultural signifiers of citizenship, in a context where full social citizenship is limited to a few. Private individuals and (mostly small-scale) businesses fill the substantial gap.

Why would this be the case? In theory, the protection of private property and the provision of public services is a key source of legitimacy for modern states. Conventional wisdom suggests that both are predicated on the ability and responsibility of governments to enable "development": to facilitate market-led economic growth, to raise standards of living, and to ensure the health and welfare of populations and the resources on which they depend. In actuality, the promise to provide public services arises for a combination of altruistic and instrumental reasons: the need to ensure a productive workforce, the assertion of political legitimacy, and reference to ethical imperatives that may (or may not) be framed in the language of human rights.

In fulfilling these expectations, the relationship between individuals and the developmental state comes to be characterized by the administration of populations by government bureaucracies enacting policy, rather than the active participation of citizens in politics that typifies the polis of Western political theory. The broadening scope of state interventions to ensure the health of populations, resources, and the economy, predicated on unprecedented techniques of surveillance and statistical representation, is what Foucault referred to as the "governmentalization" of the state.[7] Of course, the process of governmentalization is historically and geographically contingent, inflected in some instances with colonial power relations, and later shaped by relationships with development banks and aid agencies, which have devoted significant financial resources to the creation and expansion of state agencies and departments charged with the provision of public services. Nonetheless, argue scholars like Anna Stoler, the process of governmentalization in many developing countries is necessarily incomplete.[8]

Why would this be the case? Chatterjee argues that the governmentalization of the modern state poses a conundrum for developmental states, which are typically unable—for both fiscal and managerial reasons—to extend public services to the entire population.[9] These governments accordingly circumscribe such systems to wealthy elites. This pattern of allocation arises in part because postcolonial states are constrained by the patterns and policies inherited from colonial regimes. Ethnographic categories dating from the colonial period, for example, continue in some instances to define the parameters of political demands and policymaking. In the case of infrastructure-dependent utility services, exclusion from access to public services was literally hardwired into the network. Access to goods and services is usually incomplete and is thus often a locus of political struggle.

In the absence of government services, many alternative strategies of services provision emerge. Some of these involve illegally tapping into existing networks, as with the "midnight plumbers" and "comrade electricians" operating in South Africa's townships. Unregulated private alternatives spring up, usually operating on a small scale and using artisanal technologies. Some of these services may be provided by NGOs, whereas others are provided by individuals or small businesses, on a for-profit basis. Sometimes, communities come together to provide themselves with the services that their governments are unable or unwilling to provide: cooperative water-supply systems and sewerage networks, or self-built housing. All of these are "private" (in the sense of nongovernmental) strategies for providing services for and by members of the "public."

The Private Provision of Public Goods

The preceding discussion challenges our definitions of the terms "public" and "private." Policy debate tends to rely on the assumptions of utilitarian liberalism, in which the distinction between public and private equates with that between governmental and nongovernmental. Conventionally, the latter category is held to be synonymous with market (that is, for-profit) activity, although recent debates, particularly over the existence and importance of "social capital," have extended this term to include not-for-profit groups such as NGOs. From this perspective, the distinction between "public" or state authority and "private" individual activity is assumed to be clear. Debate over the scope of state and market activity is thus essentially a debate over jurisdiction, adjudicated via utilitarian standards of performance. Reducing the debate to one of technical terms or managerial strategies is, however, misleading. Privatization is a process of expansion of the market that creates new property relations and in so doing generates new society-nature relations.[10] As such, it is inherently ideological and not solely technical.

Clearly, we need to rethink our conventional usage of the terms "public" and "private," and subsequently reconsider what the process of privatization entails. Water is frequently termed, in conventional debates, a public good. This term is often used imprecisely; it sometimes means that water is nonsubstitutable, or that it is essential for life, or that its supply has critical public-health implications that have the potential to impact the entire population. In some way, people who use this term feel that it is of collective importance—hence their invocation of the term "public."

From the perspective of conventional economics, however, water is not a true public good, defined as something that is nonrivalrous and nonexcludable (street lighting and national defense are the classic examples). The consumption of water (at least to some degree) is rivalrous. Accordingly, economists and other social scientists define water as a common-pool resource, from which it is difficult to exclude access, but the consumption of which by one individual can reduce the benefits for others.

I return to the question of common-pool resources in chapter 6. Here, let me simply emphasize the confusions that often arise in popular debate, where water is frequently framed as a "public" good because of a desire to resist the characterization of water as an "economic" good. This argument is difficult to resolve. On the one hand, water has economic value, is privately owned in much of the world, and is a key economic input to a range of activities. On the other hand, water is not a true commodity, standardized and traded in bulk (except, perhaps, bottled water); indeed, converting water into a commodity is, as I have argued elsewhere, fraught with difficulty.[11] Moreover, water has a range of religious, spiritual, and environmental dimensions that suggest the treatment of water as an economic good is undesirable. Regardless of one's views of whether water *should* be an economic good, it is clear that its status is ambivalent.

This reinforces my earlier point: that the terms "public" and "private" are of little use in explaining the complex reality of water supply in urban areas. The preceding discussion suggests that three models of water supply governance can be found in cities (table 1.2). Not all of these models exist everywhere, nor are they equally dominant at any one point in time. In peri-urban and rural areas, for example, communities construct and manage artisanal (but sometimes highly sophisticated) multipurpose water-supply networks with little or no interaction with government. Large-scale hydraulic networks tend to cluster in denser urban areas; these may be managed by governments or by private companies. But here, too, community water supply exists, in the interstices of the urban fabric, in so-called informal settlements, to which networks frequently do not extend.

The ideal types presented in tables 1.1 and 1.2 are convenient fictions. In fact, many governments have created hybrid management models. Some have chosen, for example, to retain ownership while corporatizing water services, as in the Netherlands. In France, private sector management of municipally owned water-supply infrastructure via long-term management contracts is widespread. Many private companies managing water supply networks in both low- and high-income countries employ cross-subsidization schemes (similar to those typically associated with government provision) that do not charge the full cost of water to poor households. So this typology is overly tidy and thus misleading. In the next sections, I discuss each of these three approaches in turn and explore (here and in later chapters) how they overlap and interact.

The Municipal Hydraulic Paradigm

The conventional approach to managing water supply systems, in most countries around the world for much of the twentieth century, was the municipal hydraulic paradigm of water management. Predicated on an assumption of abundant water supplies, this paradigm emphasized the deployment of hydraulic technologies to meet the inevitable growth in water demands engendered by modernization. A commitment to social equity and universal provision necessitated significant government regulation, government ownership, and/or strict regulation of water resources development and water supply provision.[12] This was in line with prevailing arguments in favor of government provision of a broad range of public services, predicated on recognized advantages of government provision, both political (such as democratic accountability and a commitment to public well-being)

TABLE 1.2. *Water Supply Governance*

	Municipal hydraulic	Market environ-mentalist	Community artisanal
Primary goals	Public interest	Profit	Security of supply
Infrastructure manager	Municipal (or state) public utility	Private sector corporation	Community association or cooperative
Provision ethos	Service	Business	Livelihood
Infrastructure	Centralized	Centralized	Decentralized
User identity	Citizen	Customer	Community member ("commoner")
Accountability mechanism	Hierarchy	Contract	Community norms
Method of charging	Nonmetered	Metered	May be metered or nonmetered
Key performance incentives	Public policy goals Voter/taxpayer opinion	Price signals (share prices or bond ratings) Customer opinion	Livelihood demands Community approval
Key performance sanctions	Political process (via elections, public policy) Litigation/govern-ment appeals process	Financial loss Takeover Litigation	Exclusion of members
Raw water pricing	Subsidized or free	Market value	Varies (often "labor in-kind" equivalent)
Water supply pricing principle	Social equity (ability to pay principle)	Economic equity (benefit principle)	Community cohesion (social acceptability principle)
Water property rights	Public	Private	Common property

and technical (the relative availability of fiscal resources and the ability to collect and synthesize complex information).

Government dominance emerged after a period of experimentation with both public and private provision in the nineteenth century. In some cities, private corporations built and operated the first water-supply networks. Private companies operated in cities like Boston, New York, London, Paris, Buenos Aires, and Seville, typically supplying water to wealthier neighborhoods; the poor had to rely on public faucets, wells, rivers, and, in the most desperate cases, stolen water. The apparent inability of the private sector to finance universal provision led the government to take over the business of water supply infrastructure. In places where private companies continued to operate—as they did to a limited extent in France, England, and Spain—they were tightly regulated. Private water companies in the UK, for example, had dividends capped and were required to reinvest any remaining profits in the water supply business.

Another set of arguments in favor of the municipal hydraulic paradigm was predicated on the need of modernizing economies for large-scale mobilization of water resources.[13] Governments owned and operated most water supply systems, particularly in industrialized countries and urban areas. With the aim of providing universal access and safeguarding public health, governments created public utilities that owned the infrastructure and in most cases provided services to consumers on a subsidized basis. Water was regarded as a public service, often run at the municipal level, and was frequently not metered. Where private companies continued to operate, their activities were strictly regulated. Environmental effects of water abstraction and wastewater disposal were, in general, given little thought; the drive to supply water dominated policy and management norms.

This municipal hydraulic paradigm occurred in tandem with large-scale water resources projects. The Hoover Dam, completed in 1936 at the height of the Depression, was emblematic of a government-led drive for mastery over natural resources that was to fuel economic growth. Developing countries followed suit, in search of hydropower that would underpin the "big push" to irrigate, industrialize, and develop. Valued as symbolic infrastructure, dams (like roads) represented the sinews of the nation-state. As a means of territorializing government power, water supply infrastructure had great political as well as economic significance.

Government involvement was further justified by the socioeconomic implications of supplying water through purpose-built, dedicated infrastructure. Water is expensive to transport relative to value per unit volume, but requires large-scale capital investments in infrastructure networks (or large "sunk costs"), such that duplication of the network is inefficient. A single firm using one network is able to supply a service at a lower total cost than two or more firms using separate networks. This "subadditivity of cost" acts as an effective barrier to market entry. Water supply is thus highly susceptible to monopolistic control and is open to the abuse of monopoly power, which often leads to overcharging of consumers and suboptimal production levels. Monopolies may earn profits well in excess of the opportunity cost of capital; the putative advantages of laissez-faire thus disappear if an industry becomes monopolized.

Accordingly, after a period of experimentation with laissez-faire private provision in countries such as England, France, and the United States, the perceived failings of monopoly (overcharging, a tendency to "cherry-pick" wealthier areas and consumers, and an inability to deal with sewage and water pollution) were often cited as justifications for taking the water industry under public ownership, or subjecting it to strict regulation.

The municipal hydraulic model of network water-supply provision thus, to some extent, represented a response to experiences with private provision of water supply in the nineteenth century.[14] But even where private sector provision had not been initiated, the arguments in favor of government involvement were increasingly compelling. In essence, throughout much of the twentieth century, water was viewed as a strategic resource—like steel or coal, it was considered essential to a nation's development. As an essential lubricant of urbanization, agriculture modernization, and industrialization, water was a critical factor of production in industrialized societies; security of supply and planning for growth were key goals. This implied the need for large-scale infrastructure projects to regulate and supply water. Given the long lead times and scale of investment in water resources and supply infrastructure, government funding was often assumed to be necessary. Moreover, drinking water supply was conceived of as a public good, a necessary precondition to participation in public life, and a material emblem of citizenship.

In economic terms, the high capital costs of water supply development projects and the monopoly characteristics of water supply networks were used to justify government involvement. In most industrialized countries and urban areas, governments dominated the business of water supply throughout much of the twentieth century. Where governments set up corporations to run water supply systems (as in the Netherlands) these tended to be nonprofit. The justification for government control of water supply systems thus rested on twinned economic and ethical arguments.

Strong normative claims were also made in favor of government involvement. Justifications, particularly in the twentieth century, included the symbolic and cultural importance of water as a (partially) nonsubstitutable resource essential for life; its strategic political and territorial importance; the intense conflicts that arise over the use of a flow resource required to fulfill multiple functions (agricultural, industrial, drinking-water, ecosystem health); and the need in industrialized, urbanized societies to mobilize large volumes—invariably at a high cost relative to the economic value generated.

Sufficient supplies of safe water enabled a basic, dignified living standard and facilitated social inclusion; as such, access to sufficient amounts of clean water came to be viewed as a precondition to participation in public life.

Yet by the late twentieth century, this municipal hydraulic paradigm had come under attack. Greater awareness had also emerged about the (still hotly debated) effects and (often unquantified) costs of hydraulic development, particularly large dams: namely, extirpation of species (particularly fish), displacement of communities, flooding of cultural sites, contamination of water sources, disruption of ecological processes, and environmental degradation.[15] With threats to human health from water-borne diseases such as typhoid under control, concern began to focus on nonpoint sources of pollution and other contaminants, as evidenced by clean water legislation in the United States and similar legislation on water quality in the European Union.[16] This concern was heightened by the growing realization that the postwar economic boom had obscured systematic deterioration of water supply infrastructure in many countries.[17] In a period of fiscal and ideological government crises, the inability (or unwillingness) of governments to finance infrastructure combined with public environmental concern and technological innovations to increase the appeal of alternative approaches to water management.[18] The municipal hydraulic paradigm thus faced a multidimensional challenge: ecological, cultural, ideological, and socioeconomic.[19]

In short, the failings of the municipal hydraulic model attracted significant criticism over time. Of particular concern was the supply-led focus, in which it was presumed that economic growth necessarily implied increasing consumption of water resources (leading in some cases to "white elephant" projects). Criticism was also motivated by the systematic exclusion of environmental considerations (such as effects on aquatic ecosystems). Finally, the phenomenon of urban water scarcity—in which rapid urbanization, limited public finance, and the informalization of economies overwhelmed municipal governments' abilities to supply networked water services—lent urgency to reform agendas. This set the stage for the emergence of market environmentalism.

Market Environmentalism and the Private Utility Model

If the dam-as-icon symbolizes the modernist impulse underlying the municipal hydraulic mode of regulation,[20] the tradable, artificial wetland is the archetype of a market environmentalist approach to water resources management.[21] Restoration ecology and economics displace hydraulic engineering as dominant paradigms for interventions in hydrological landscapes.

"Nurturing nature" holds out the promise of enhanced environmental value (amenity and aesthetic), while orchestrating our interactions with the environment to more efficiently provide functions previously supplied through infrastructure-intensive developments (e.g., treating marshes as water purification systems and using biological rather than industrial processes).[22]

These dimensions of market environmentalism vary in importance in different geographical contexts. In lower- and middle-income countries, private sector participation in water supply in urban areas has increased dramatically over the past two decades, and the irrigation sector has been the target of pricing and market reforms in rural areas. In high-income countries, greater priority has been given to private sector management of urban water supply networks, ecological restoration, and water quality improvement.[23]

Market environmentalism as applied to water is not monolithic; it varies considerably with jurisdiction, socioeconomic framework, and cultural setting. Yet one can identify several common characteristics (table 1.3).[24] Water conservation, for instance, is prioritized, and challenges the assumed link between economic growth and increased water use. There is a growing emphasis on valuing the environment and incorporating environmental values into policy. Environmental protection, remediation, and restoration assume increasing importance as goals (and constraints) of water supply management. In pricing, the priority moves from access or equity toward economic efficiency. In some instances, private-property rights and tradable markets for water may be established; more often, government regulation applies market-based or market-simulating techniques. Overall, the private sector, as both owner and manager of the infrastructure, tends to play a relatively more important role than the government.

Another key component of market environmentalism is a set of changes to governance that are conventionally labeled "commercialization," in which markets and market norms are applied to water supply management. Commercialization frequently introduces metering and associated changes in water rates. The principles of full-cost pricing are usually applied, according to which prices should reflect the full cost of infrastructure and maintenance, and consumers should pay for what they use. This is in contrast to subsidized pricing and social equity principles (consumers should pay according to their ability and what they can afford) that tend to shape public utility systems.

Technically, the focus shifts from creating new sources of supply to managing demand through a variety of techniques (conservation, new water-saving technologies), alternative supplies (gray water, reclaimed waste

TABLE 1.3. *Market Environmentalism: Examples from the Water Supply Sector*

	Target	Strategy	Example
Privatization	Asset ownership	Divestiture (also termed "privatization" or "asset sale")	Sale of water supply infrastructure to private sector (England and Wales [Bakker 2004])
	Resource property rights	Marketization	Introduction of a water market (Chile [Bauer 1998])
Commercialization	Business model (organizational structure)	Corporatization	Conversion of business model for municipal water supply: from local government department to a publicly owned corporation (Amsterdam [Blokland, Braadbaart, and Schwartz 1999])
Liberalization	Utility network management	Private sector participation (outsourcing contracts)	French municipal outsourcing of water supply system management to private companies (Lorrain 1997)
	Performance incentives/ sanctions	Commercialization	Introduction of commercial principles (e.g., full cost recovery) in water management (South Africa [McDonald and Ruiters 2005])
	Regulatory frameworks	Deregulation and delegation to private actors	Delegation of direct state oversight of water quality to private actors (Ontario, Canada [Prudham 2004])
	Governance	Decentralization	Devolving water-quality monitoring to lower orders of government or individual water users (Babon River, Indonesia [Susilowati and Budiata 2003])

water, desalinated water, recycled water), metering, new tariff structures, and educating consumers in a new "ethic" of water use.[25] Economic equity (the "benefit" or "willingness-to-pay" principle) displaces social equity (the "ability-to-pay" principle) in water pricing.[26] Consumer access is legitimated not by a citizen's entitlement to water as a service but by a customer's purchase of water as a quasi commodity. Thus, a reconfiguration of the hydro-social contract between users and their environment is required. Market environmentalism implies that consumers paying per unit volume at cost-reflective prices will use water more efficiently than unmetered households or farmers accustomed to treating water as a free service.

Three processes constitute the core of market environmentalism: the privatization of resources, the commercialization of environmental management, and the liberalization of governance. *Privatization* entails the transfer of ownership or management of resources to private persons (individuals or companies), and the creation of new markets for previously uncommodified resources—for example, water privatization and the patenting of genetic material unearthed through bioprospecting. *Commercialization* entails the incorporation of market actors and market-oriented institutions (rules, norms, and customs) in resource management; tradable quotas in fisheries and debt-for-nature swaps are examples.

The *liberalization* of governance—the third core process of market environmentalism—entails the retreat of governments from decision making in the classic move from "government to governance" via deregulation, devolution, and decentralization. More precisely, liberalization entails selective deregulation and reregulation designed to encourage competition and reduce the role of government agencies in regulation and decision making more broadly. Frequently (although not always), this is allied with a marketization of governance, in which private sector norms and market principles are introduced into the civil service (sometimes referred to as "new public management"), government administrative functions are contracted out to nongovernmental actors, and market-based allocation and performance measures and incentive structures are introduced. The European Union's Water Framework Directive provides an example of the simultaneous marketization and rescaling of environmental governance, as it mandates full-cost pricing, the creation of local watershed councils, cost-benefit analyses, and collaborative decision-making processes. Market environmentalism may thus incorporate privatization but is much broader in scope. Indeed, in some instances it may not entail privatization at all; markets can be introduced and decision-making principles can be commercialized without a formal transfer of ownership.

The specific nature of these reforms varies across countries and regions. But the general trend, as we shall explore in chapter 3, is clear: at the beginning of the 1980s, government management of water supply was an exception rather than the rule. Two decades later, thousands of water supply systems in developing countries are managed by a handful of private multinational companies eagerly pursuing growth in a multibillion-dollar global market. In doing so, they displace not only government managers but also (although more rarely) community water-supply systems.

Community-Run Water Supply Systems

Conventional thinking in development circles holds that community-run water-supply systems represent a distinct break from the two models we have just described. Community has been a dominant theme in development discourse for the past half century, and it has been imbued with multiple definitions with various claims to legitimacy: from indigenous (at times essentialized) invocations of ancestral community, to neoinstitutionalist framings, to (post)colonial constructions of community as nation.[27] Water use and governance practices within communities are of a startling diversity, as an extensive academic literature demonstrates. This ranges (to mention just a few) from documentation of traditional water-use practices (such as Gilbert White's work on water carrying in West Africa) to the institutional dimensions of common-property water management regimes (such as Ostrom's classic text on Spanish *huerta* irrigation systems).[28] More recently, academics have also begun exploring the links between water use practices and technologies, on the one hand, and community governance and cultural constructions of water's biophysical and symbolic dimensions, on the other. Lyla Mehta's work, for example, documents how water scarcity in India is mediated by inequitable power relations.[29]

A central theme running through much of this work is the autarkic (in the sense of self-sufficient) nature of community-managed water systems. This is particularly the case with rural or peri-urban water supply, where irrigation uses are of primary importance, and in which multiple uses of water (for drinking, bathing, and other household purposes) coexist. Local in nature, using small-scale (or "appropriate") technology, community systems mobilize local labor and community participation, and although they are articulated with formal structures of government, they almost invariably vest governance at the community level. Often these community systems are reflective of worldviews that differ vastly from the Western engineering mind-set that conventionally governs large-scale, industrialized water-supply networks. Community water supply, from this perspective, is implicated in cultural politics and in the social and political identities of water users.[30] Community irrigation provision in the Andes is one well-known example. Cultural practices, religious beliefs, and local customs shape rural irrigation by Andean communities, in which land and water use, spirituality, and community relationships are intertwined.[31]

Rendering Technical

In many instances, this diversity and autonomy of community water governance practices poses a challenge both for water management agencies and for the engineering expertise on which they typically depend. This is, of course, not unique to water. As James Scott has argued, modern systems of centralized management necessarily entail simplification, standardization, and homogenization. Scott argues that standardization usually involves the imposition of a generic set of norms, rules, and associated technological knowledge ("techne": the logic of bureaucracy and hierarchy in modern industrialized societies) that often results in material and psychological violence to nature and humans alike.[32] He posits *mētis* in opposition to techne, where *mētis* refers to local, embedded, context-sensitive, community-based knowledge—which is inevitably heterogeneous and unruly, hence difficult for large organizations (whether public or private) to incorporate and adjudicate in their drive to optimize resource production. Jacques Ellul makes a similar argument regarding "technique": the calculus of standardization, rationalization, and alienation characteristic of large-scale, centralized, corporatist management systems.[33] For example, dependent on systematized, decontextualized knowledge, the managers of modern hydraulic systems sometimes assume that lack of access to water is a purely technical problem, and that purely technical solutions can be found. This process of "rendering technical," as Tania Li terms it, effectively forecloses consideration of other factors, be they political or cultural.[34]

Following Scott's argument, we might assert that governments have frequently aimed to curtail, reshape, or even eradicate the local diversity found in community water-supply management systems, displacing community knowledge of water supply systems with the techne of hydraulic engineering, accounting, and water economics. This does not occur without resistance: as Rutgerd Boelens has explored in the Andes, community struggles to defend water access are frequently framed as a defense of the legitimacy of community authority and community water rights as opposed to those of external actors (whether governmental or private).[35]

The policy literature on community water supply in urban areas often tends to overlook these issues. Relatively little emphasis is placed on the ethnographic dimensions of water supply: cultural practices, religious beliefs, and customs are less frequently studied and often dismissed as anachronistic.[36] Documentation of nonnetworked water-gathering practices is relatively rare.[37] Nonnetworked provision, where it exists, is rarely framed in terms of resourcefulness or self-sufficiency. Rather, nonnetworked solutions are often framed as temporary and of dubious quality. The network is assumed to be the norm, and the continued existence within urban areas of nonnetworked technologies of water supply provision controlled by communities is often portrayed as problematic. For example, the acknowledgment of community resistance to state control, which is evident in the literature on rural water supply, is much more rarely evident in the literature on urban communities without water access, who are typically characterized as the "thirsty poor."

I return to this point about portrayals of the thirsty poor in chapter 4. Here, I want to emphasize the ways in which the concept of community intersects with the debate over public and private water supply. On the one hand, communities fit uneasily with the corporate models of governance associated with industrialized, urban water supply systems. They both threaten and are threatened by modern water-management norms based on engineering (and, increasingly, economic) expertise. On the other hand, the concept of community challenges our notions of "public" and "private." As illustrated by the story of the three end users at the start of this chapter, community water-supply practices usually entail a mix of for-profit and not-for-profit strategies, which are carried out by a shifting combination of private individuals and community groups. These water use strategies are not "public" in the sense of governmental (indeed, they are usually not sanctioned by governments) and not "private" in the sense of formal corporations. But they are often (although not always) undertaken for profit. This confusion is evident in the terminology used to describe these activities in an urban context: "informal," "small-scale," or "local."

The existence of community water-supply systems in urban areas also presents us with a dilemma. Should community water-supply strategies be integrated into urban water management, or eliminated through the expansion of networks? This question is particularly pertinent given two trends: the rediscovery of community by both proponents and opponents of private sector participation in urban water supply network management, and the decentralization of water governance to local government or watershed scales in many countries.

I will return to the issue of community water management, and to the dilemma it poses, in chapters 6 and 7. Here, I will simply stress that the problem of managing resources at multiple scales—necessary in the case of water—poses problems of cooperation and coordination among large groups of users. The local nature of water abstraction (particularly groundwater) raises questions about problems of overuse. And the tight linkages between users, particularly in terms of water pollution, raise the risk of "free riding" of some users on others. The mobility of water increases the costs of measurement and information significantly and often beyond what local communities could afford (or be capable of collecting). And communities will not always have the expertise (or desire) to anticipate and manage environmental quality issues. Water, in short, is difficult for communities to sustainably manage. Some of these difficulties are resolved by the creation of networks, but these, in turn, pose new problems. In other words, like the other two models previously proposed, community water-supply management is not a panacea. In the next section, we turn to a detailed consideration of the problems posed by the three models.

GOVERNANCE FAILURE, MARKET FAILURE, STATE FAILURE

Market Failure, State Failure

Traditionally, the argument in favor of government provision has been justified by reference to the concept of "market failure." Simply put, market failure occurs when markets fail to efficiently allocate goods and services. The locus of failure is the mechanism of exchange in the pursuit of monetized private interest (usually via the market and its pricing mechanisms). From the perspective of neoclassical economics, this arises in situations that fail to correspond with standard assumptions, resulting in effects that are not (and perhaps cannot be) priced, and therefore are not accounted for by economic actors.[38] We might as easily use the term "exchange failure" (although this is not used in the literature, and for convenience I continue to use the term "market failure" throughout the rest of the book).

How does this apply to water? Water is heavy to transport relative to value per unit volume, requiring large-scale capital investments in infrastructure networks, which act as an effective barrier to market entry. Water supply is thus highly susceptible to monopolistic control (a condition termed "natural" monopoly by economists). Commercializing water utilities is as a consequence invariably fraught with difficulty, to a much greater extent than for other network utility services such as telecommunications, gas, and electricity.

Moreover, lack of access to water has important (and potentially disastrous) consequences for hygiene and public health; pollution of water sources by one user can quickly affect others. These "externalities," together with the tendency of private companies to fail to extend coverage to the poor (as a result of both the tendency to cherry-pick profitable neighborhoods and classes of consumers, and the high prices and poor services resulting in a situation of natural monopoly), were two of the most important justifications for bringing water supply under the control of the government during the twentieth century. The scale and significance of these externalities—particularly acute in the large urban centers of the nineteenth century ravaged by water-borne diseases such as cholera and typhoid—was an important justification for producing water as a quasi-public good through the provision of universal networks. Like national defense or street lighting, all members of society tended to benefit from the health improvements resulting from clean water supplies, regardless of whether they paid for the service or not. And since people were believed to be for the most part unaware of the benefits of improved water supply (and particularly sanitation), water was also defined as a "merit good," which, if provided solely by the private sector at market rates, people were likely to underconsume, again providing a strong justification for government provision.

Proponents of privatization, and of market environmentalism more generally, counter these arguments with the concept of "state failure." Government failure, as it sometimes also called, is the thesis that governments are less productive, efficient, and effective than markets (at times conjoined with the notion, derived from neoconservative or anarchist perspectives, that governments are despotic and inimical to freedom). In a slightly different formulation, Bob Jessop defines state failure as "a failure to secure substantive collective goals based on political divination of the public interest."[39] In the water sector, the state-failure argument comes into play to support claims that governments manage water poorly, both through failing to supply safe water to the poor and through allowing severe environmental degradation of water ecosystems in the pursuit of goals such as modernization, industrialization, and urbanization.

The state-failure critique of public water provision echoed the widespread neoliberal turn in public policy and development policy of the 1980s, shaped by public choice theory and the new institutional economics. Governments, acting as both owners/operators and regulators, are subject to the "poacher-gamekeeper" problem; without incentives to penalize themselves, performance would deteriorate, particularly with respect to environmental externalities. Rent seeking by officials might lead, for example, to overstaffing or "ghost" employees, or to bribery and corruption among staff. Without appropriate incentives, public water companies would be likely to operate inefficiently. And capital-intensive hydraulic works were particularly susceptible to pork-barrel politics, leading to overdesigned "white elephant" projects that would require heavy subsidies from government for both capital and ongoing operational costs. "State failure," in other words, refers to the shortcomings of large-scale hierarchies and the associated decision-making and incentive mechanisms that we typically associate with governments; another way of phrasing this would be "hierarchy failure." The implication follows that governments should cede regulatory control to markets, while relinquishing direct management of services and ownership of the assets employed to provide those services.

These distinctions are not so tidy in practice. The failures of the private sector resemble, in some cases remarkably, those putatively due to governments: an inability to incorporate environmental externalities, to source adequate amounts of finance, or to extend water in a significant way to the poor. Indeed, many of the criticisms directed at either governments or private firms could equally be applied to the opposing target. James Scott's blistering critique of government management of resources (discussed earlier) is a good example.

Governance Failure

Critiques of market or government provision are often, I would argue, misdirected attacks on Ellul's "technique" (or Scott's techne). Critics often (incorrectly) assume that the introduction of techne can be ascribed to either government or private actors, and fail to realize (or at least to admit) that many "techniques" are common to both. If this is true, how might we frame the failures of public and private water management in more general terms?

Relatively little of the debate over private sector participation in urban water management has been framed in this way. Here, it is helpful to draw on debates that expand our concept of governance beyond the technical definition given above. In its classic formulation, the term refers to the problem of coordination in societies comprised of heterogeneous interests and capacities. From the classic definition as "the art of steering society"[40] we might tease out a more elaborate explanation that includes decision-making processes, associated institutions (in the sociological sense of laws, rules, norms, and customs), and the identities and subjectivities in which they are embedded. Governance may, in other words, be defined as a process of decision making that is structured by institutions (laws, rules, norms, and customs) and shaped by ideological preferences. In the language of policymakers, governance may be defined as the range of political, organizational, and administrative processes through which stakeholders (including citizens and interest groups) articulate their interests, exercise their legal rights, take decisions, meet their obligations, and mediate their differences.[41]

The term "governance" has risen in prominence in recent decades as formal government authority has increasingly been supplemented or supplanted by a reliance on informal authority; roles previously allocated to governments are now (controversially) categorized as more generic social activities carried out either by political institutions or by other actors.[42] This is often characterized as a move from "government towards governance."[43] More accurately, we might refer to this process as the "hollowing out of the government," which often entails a rescaling of governance to sub- and supranational scales.[44] For some, this is a positive development. Here, "governance" is often used in a normative sense, naively celebrating the rise of nongovernment actors without questioning the reasons for the "retreat of the government" or considering its implications. This view merits concern, particularly as it glosses over the implications of decentralization and delegation.[45]

What might the concept of governance contribute to debates over state and market failure? The term "governance failure" might be useful in describing how the institutional dimensions of water management and decision making do not effectively take into account the needs of all citizens.[46] Phrased differently, governance failure occurs when there is a mismatch in decision making between citizenship and associated political rights, on the one hand, and institutional and cultural practices on the other. This phenomenon is not specific, of course, to water. Jessop defines governance failure, more generally, as a failure of the "self-reflexive deliberation and self-organization that occurs in networks, public-private partnerships, corporatism, private interest government, regulated deregulation, and so on."[47] As such forms of governance become more widespread (as is the case, for example, under market environmentalism), the question of governance failure becomes more acute.

Here, it is important to note that state, market, and governance failures are not associated in any direct, linear fashion with the models of water supply or actors—governments, private companies, and communities (or networks)—that we have described. To the extent that all of our real-world organizations are a mix of models (a mix, for example, of the hierarchies and exchange practices typically associated with governments and markets, respectively), they will exhibit a range of what I have here termed "failures." A key point to consider is that some forms of public services might more effectively and efficiently be associated with one model (market, command-and-control, decentralized network) than another; and, equally, that governance failure might be more acute with respect to some forms of provision than others. (See table 1.4.)

TABLE 1.4. *State, Market, and Governance Failure*

	Examples
State failure	1. Rent seeking (by officials and politicians)
	2. Unincorporated externalities
	3. Poacher-gamekeeper problem (if both supplier and regulator are public)
	4. Regulatory capture (if supplier is private)
Market failure	1. Imperfect competition
	2. Asymmetric information
	3. Unincorporated externalities
	4. Public good
Governance failure	1. Political disenfranchisement (e.g., lack of "voice" on the part of poor households)
	2. Culture of decision making (e.g., elite-focused, top-down)
	3. Absence of consumer entitlements to basic services (e.g., lack of universal service requirement)

Of course, these three concepts are, to a degree, caricatures. But, taken together, they are a powerful means by which we can reframe the problem of urban water supply. To give a simple example, take the case of water supply in Dhaka, Bangladesh.[48] Theoretically, all residents of the city should have access to water supply, but water is in practice only available to middle-class neighborhoods, and it flows with regularity and sufficient pressure only in elite enclaves within the city. Elsewhere, the use by wealthy families of private pumps to illegally draw water from underpressurized water-supply mains implies that water supply dries up to a trickle for the majority of residents. This fragmentation of the water supply network is deliberate and was shaped by both colonial practices and the politics of successive postcolonial regimes. City residents have little means of changing this situation, for the water supplier's advisory board is made up of twelve members from the city's elite, and decisions about water supply access and availability are made behind closed doors. In 1997, a private sector company was brought in to manage the network, but this has changed little for city residents. The technological systems remain centralized and highly capital intensive; management remains highly corporatized; decision making is largely left to engineering experts. Official policy is that water supply should only be provided to legal households, with official land permits. The 3.5 million slum dwellers without these permits are not regarded as potential customers and have no officially sanctioned mechanism for influencing decision making.

They are, as Appadurai terms it, "citizens without a city,"[49] those to whom modern norms of social citizenship do not apply. Their experience of water supply suggests that it is simultaneously subject to state, market, and governance failure. Indeed, it is the simultaneous occurrence of these multiple failures that makes water such an unruly resource to manage.

This formulation suggests that decision-making structures and related institutions may contain systematic biases against certain groups—such as poor households. It also suggests that the debate over the relative merits of public versus private ownership and/or management of water supply systems has, to some degree, missed the point.[50] A focus on governance suggests that the institutions we choose for managing water—and the principles and ideologies that support these institutions—play a powerful role in shaping the governance process and in determining the likelihood of equitable, sustainable outcomes.

The definition of governance failure introduced above obviously overlaps with Sen's concept of capabilities.[51] It suggests that constraints and opportunities are experienced by individuals and shaped by their values, choices, and preferences. It allows, in other words, for the reentry of individual agency and ideology into the debate.[52] In this sense, the concept of governance failure demands that we account for the broader ideological commitments and systems of rationality that structure decision making and social relations in which water supply is embedded.

But, as outlined below, the reality of urban water supply is only imperfectly described by terms such as "governance failure," or "market failure." The categories of states and markets, or public and private, and the term "governance" as a means of capturing the articulations between the two, are unhelpful in delineating the contours of the problem. Conflating "public" with "state" provision obscures collective forms of action not mediated by states. Equally, conflating "private" with "for-profit, corporate" activity is misleading, as it mistakenly assumes that all nonstate activity is necessarily capitalist. The term "governance," in turn, allows the fiction of apolitical negotiation to be sustained and enables development-policy experts to relegate failure to the merely technical realm.

The Politics of the Governed

If the notion of governance (and governance failure) as conventionally deployed is insufficient to the task at hand, where might we turn instead? Chatterjee's distinction between "civil society" and "political society" allows us to redress the apolitical fictions sustained by terms like governance.

When considering the relation between modernity and democracy, Chatterjee argues that the concept of civil society—predicated on conventional notions of nation-states, popular sovereignty, and citizenship, via a normalizing, inclusive set of power relations—is an incomplete description of the practice of politics in most of the world. This, it should be emphasized, marks a deliberate contrast to the situation in "developed" countries, where civil society encompasses the entire population of a nation-state, and where Foucault's notion of "governmentalization" (discussed earlier) applies to the quasi totality of the population.

In cities in the South, Chatterjee argues, civil society refers to a small minority who are fully rights-bearing citizens in the sense imagined by Western political theorists. To characterize the remainder of the population, Chatterjee coins the term "political society," which refers to the majority of people who "are only tenuously, and then only ambiguously and contextually, rights-bearing citizens."[53] Government institutions approach the latter groups as "populations" (such as "the poor" or "slum dwellers") that must be simultaneously controlled and cared for. This implies a political relationship between populations (rather than rights-bearing individuals and the associations that represent them) and the state, in which governmental agencies pursue policies that attempt to provide for the security and welfare of populations beyond the scope of the institutions mandated by constitutions, laws, and formal justiciable rights.

This political struggle over what some scholars term the "right to the city" is often fraught with conflict, because although government agencies recognize that populations do have claims on the welfare programs (such as access to education, water, or housing), they frequently do not have the means to deliver those benefits.[54] As Chatterjee notes, governments are confronted with a dilemma: "To treat those claims as rights would only invite further violation of public property and civil laws," yet the legitimacy of governments is founded on the promise of extending the material entitlements of modernity, such as utility services, to all.[55] As a result, these claims are continuously negotiated between governmental obligations, on the one hand, and political expediency on the other.

In some countries, an associated response is a retreat of the elite, in which certain segments of society are excluded from the project and promise of modernization. This is, of course, particularly critical in "megacities," which play an increasingly important role in the global economy yet are beset with serious problems of environmental sustainability, including overexploitation and pollution of surface and groundwater (but also air pollution, solid waste, and other issues). The scale and extent of illegal

settlements in megacity regions exacerbates these problems, with environmental degradation and social erosion mutually reinforcing one another.
As explored in subsequent chapters, the problem of urban water is thus
simultaneously a socioeconomic and environmental issue, often obscured
by the failure to view urbanization as an ecological as well as a political-
economic process.

In many instances, this process occurs in the context of "urban primacy,"
in which most large-scale modern activities, forms of social infrastructure,
foreign-exchange-generating economic production, and centers of decision
making are made or controlled through a single major city.[56] In creating
and sustaining capital or showcase cities as political symbols and nodes of
economic activity, public resources are spent on the privileged consumption
of a small minority of urban residents; only these citizens are considered
to be political constituents of society with full entitlements to state services. Unequal access to water supply is thus not simply the outcome of
neoliberalism or vulture capitalism (the "market failure" argument), or of
incompetent or self-interested governments (the "state failure" argument).
The distinction between citizens and populations is literally embedded in
the city's infrastructure, through the interrelated production of subjectivities (subalterns versus citizens), spaces (the city and the slum), and infrastructure (the network versus the hand-dug well). This is simultaneously a
cultural as well as a political process, which underpins inequitable urban
governance.

CONCLUSION: REFRAMING URBAN WATER SUPPLY, RETHINKING PRIVATIZATION

This chapter has identified a central dilemma posed by the urban water-
supply crisis: in many cases, governments have failed to adequately deliver
water supply, but private companies have largely failed to improve the situation. Chapters 2, 3, and 4 will mobilize evidence to support this claim. As
with all generalizations, counterexamples can easily be found. Nonetheless,
I will assert that this dilemma holds true for many countries.

What are the implications? The failures of both government and the
private sector to deliver "water for all" suggest that more attention should
be focused on the ongoing dialectic of inclusion and exclusion experienced
by urban residents. The concept of governance is useful for excavating the
social power relations that underlie this phenomenon (if we are careful to
rescue the concept from the depoliticized approach to which it is prey in

policy documents). In other words, urban water—all too often treated as a technical issue—is inextricably bound up with issues of governance broadly understood to include political processes, social power, collective delibera- tion, and the mediation of competing interests.[57] In subsequent chapters— particularly chapter 4—this approach will be used to provide an alternative (although necessarily partial) explanation for the persistent failures of both public and private operators in managing urban water supply utilities.

Reframing the urban water-supply crisis thus requires us to challenge conventional assumptions about "public versus private." The strategies through which urban residents supply themselves with water (and indeed with many other essential services) are complex admixtures of quasi-public and quasi-private institutions that defy easy classification. As managers of water supply systems, governments and private companies display impor- tant similarities as well as differences. The question of which model, or blend of models, is preferred has as much to do with our ideological prefer- ences as with our understandings of the advantages and shortcomings of public, private, and community governance.

This suggests that the urban water-supply crisis stems, in part, from the inability of governments to supply the services on which their politi- cal legitimacy is predicated. Of course, this is a source of grievances and political protest in many places. Simultaneously, alternative modes of sup- plying people's basic needs proliferate, including surreptitious connections to networks offered by midnight plumbers, illegal "spaghetti networks," or "ghost customers." Yet these actors do not always make choices that seem aligned with the development mantra of "water for all." This suggests that the problem is one of collective action with political, economic, and eco- logical dimensions, in which not only governments and private companies but also communities are actors. In simple terms, the urban water crisis is partly an outcome of poor governance and is characterized by what we might term "governance failures." However, we would be mistaken if we were to approach governance in narrowly institutionalist terms (e.g., rules or decision-making processes): my reference to Chatterjee is intended to imply a broader view, in which governance is shaped by cultural norms and practices.

This also suggests that we need to rethink conventional framings of the process of water privatization. In high-income countries, corporatized, large-scale, centralized, and capital- and technology-intensive forms of water management (both public and private) have largely displaced the community management of water as a common-pool resource. For main- stream economists, this is a legitimate process, which is usually understood to take the form of a noncoercive exchange between buyer and seller. Political economists assert, in contrast, that privatization is often coercive, and an act of dispossession. But in both cases, privatization is understood to imply a transfer of control (and, more rarely, ownership) of industrial water-supply network assets, together with their client base, from governments to private corporations. However, these analyses overlook an important point (which has been central to this chapter): the creation of large-scale, integrated, modern technical networks is far from complete in lower- and middle-income countries, where community control (small scale, decentralized, and reliant on alternative or artisanal technologies) remains widespread. The urban poor are already dispossessed, if by this term we mean disconnected or excluded from access to the network.

Subsequent chapters develop these arguments further. The analysis they present is predicated on the assumption that the urban water supply dilemma will not be solved solely by better "models" of public or private provision, or by better governance institutions. To achieve meaningful access to water while protecting the hydrological environments on which we all depend, critical examination, reform, or even replacement of the models presented in this chapter is necessary if we are to formulate adequate responses to the failure of both conventional government and private sector models to satisfactorily provide "water for all." Subsequent chapters in part 1 focus on a critical examination of our dominant models of water supply, and the issue of reform is taken up in part 2.

2

MATERIAL EMBLEMS OF CITIZENSHIP

CREATING PUBLIC WATER

The distinction between "public" and "private" which dominates the water privatization debate is, in fact, a relatively recent invention. This chapter presents a historical perspective on urban water supply, beginning with the "industrialization" of water supply that took place over the course of the late nineteenth and twentieth centuries. In the words of historian Jean-Pierre Goubert, water was literally conquered through new hydraulic technologies, which supported rapidly rising rates of consumption in both developed and developing countries.[1] Water supply infrastructure—and associated household technologies dependent on running water in large quantities—became widespread, and were associated with civic life; hence, as I explore below, these merit the label of "material emblems of citizenship." These large-scale water supply systems were associated with new scientific and cultural conceptions of the multiple roles of water—aesthetic, economic, and social—within modernity. At the same time, new definitions of "public" and "private" emerged; their application to water supply was dependent on a set of techniques and technologies discussed in the previous chapter.

Taken together, these technologies and ideologies constitute the basis of the "municipal hydraulic" paradigm (introduced in chapter 1). This paradigm was not unique to water; it applied, with some variations, to all universalized, integrated utility networks (e.g., gas, telephones, and electricity) that underpinned infrastructure development—particularly in urban areas—throughout the second half of the twentieth century.[2] The history of this paradigm, and the challenges to which it was subject, particularly in developing countries, are the focus of this chapter. In short, this chapter builds on the analysis presented in chapter 1, as it documents the emergence of the "municipal hydraulic" paradigm and associated "state failures" in urban water supply in the twentieth century. This is important for those interested in issues of privatization, because the performance of the municipal hydraulic paradigm (and discursive constructions of this paradigm as a "failure") was a key justification for the emergence of private sector participation in the 1990s.

This, in turn, requires an understanding of the approach to water supply within the "development" sector over the latter half of the twentieth century. Accordingly, I devote a case study to the water-related development policies of the World Bank and its affiliates (known as the World Bank Group, but labeled for convenience as "the Bank" in this chapter). With its large water-related lending portfolio, the Bank set the agenda for the development community, supporting the introduction of the municipal hydraulic paradigm in cities around the world. The analysis of the evolution of the World Bank's policies (based on archival research) suggests, surprisingly, that the inclusion of water supply in the Bank's portfolio was neither uncontested nor straightforward. Covering the period from the 1940s to the present, the analysis highlights critical difficulties—ecological, political, socioeconomic—that the Bank's water projects encountered, and articulates this analysis with the emergence of "market environmentalism"—and hence privatization—in the Bank's water policy in the 1990s. In so doing, it identifies some devastating flaws in the Bank's approach to water supply lending and in the expertise it brought to bear on questions of water supply. This lays the ground for an understanding of the origins of private sector participation in water supply, which is the focus of chapter 3.

This history is important, as it allows us to avoid a purely ideological set of explanations for the rise of privatization. Rather, it acknowledges that material conditions of urban water supply services (and indeed other infrastructure services) were, by the 1980s, severely degraded in many cities around the world. This suggests that the municipal hydraulic paradigm, as applied to urban water supply, had failed in reaching its stated goals, in part due, as we shall see, to an interrelated set of governance failures. The current debate over privatization, to put it simply, has its historical roots in a crisis in water supply systems related to broader issues of development, governance, and democracy. This alternative explanation in turn provides the basis for the reframing of urban water supply presented in subsequent chapters (particularly chapter 4).

Here, a comment on the role of communities in water supply is merited. The introduction of the municipal hydraulic paradigm often entailed (and today still entails) the implementation of corporate forms of water supply ownership, production, and distribution and the concomitant displacement of community-controlled water-supply systems. This process is, however, by no means complete or uncontested.[3] Some communities (often in rural and peri-urban areas, in "marginal" areas of developing countries where community control of water supply remains central to subsistence) reject the encroachment of states and private companies, viewing the imposition of modernized systems as akin to processes of "enclosure."[4] Subsequent chapters return to these debates and to the question of community-controlled water-supply systems; for the moment, a simple caveat will suffice: that the municipal hydraulic paradigm discussed in this chapter is neither complete nor hegemonic. It is, however, the dominant paradigm for urban water supply around the world and has been the primary target of proponents of privatization—and hence merits our attention.

MODERNIZING WATER

The experience of modernity is intimately, viscerally associated with water. In the West, the role of water as a resource, and aesthetic and cultural views of its place in society, changed dramatically during the nineteenth century: water-use practices became a source of sensual pleasure, the object of new, water-intensive personal hygiene routines, and a marker of civilization.[5] In the twentieth century, dams and reservoirs were symbolic of the twin projects of modernization and nation-building.[6] Economic growth was predicated on industrialization, urbanization, and agricultural intensification. This, in turn, required a secure, continuous supply of large amounts of relatively high-quality water, which dams and newly "regulated" rivers could provide.

The domestic counterpoints of the dam are the washbasin and the toilet, symbolic of the sensual and aesthetic role played by water in urbanization and cultural modernization. The bourgeois residents of nineteenth-century European cities, for example, celebrated hygiene as a moral virtue. Strong links were made by an alliance of the medical establishment and social reformers between clean material surroundings and moral rectitude.[7] These reformers linked physical and moral "cleanliness," espousing a vision of a moral order in which "physical well-being, the preservation of the family and the maintenance of morality" were essential to the "security of society."[8] They encouraged the lavish use of water in newly interiorized, private spaces such as the bathroom and boudoir, with (then) new technologies such as the flush toilet.[9] These technologies embodied a set of assumptions about water's renewability, abundance, and "self-cleaning" abilities. Water supply networks depended on the practice of using surface watercourses as the "ultimate sink" for pollution, diluting and carrying away wastes from the factories and living spaces of the cities.[10] These were intended to support the provision of adequate amounts of affordable, safe water to all citizens.

For these reformers, water supply was a material expression of political inclusion. Citizenship, they argued, must be conceived not only in terms of political representation but also services provision. This was, in some ways, a radical departure from earlier visions of citizenship, and these types of arguments underpinned, in part, the massive growth in the size of welfare states in the twentieth century.

Reformers' advocacy of universal provision by governments was underpinned by the claim that access to water and disposal of wastes is critical not only for preventing disease but also for a minimum level of dignity to which all citizens have a right. Whereas dams symbolized the technological prowess of modernizing nation-states, water supply networks became the emblems of their civility. Access to water came to be depicted as a precursor for the production of modern citizens.

This implied much greater volumes of water use than in the past. Residents of medieval cities were parsimonious in comparison to today's urban elite. In medieval cities, tap water, indoor toilets, and sewage-disposal networks were absent. Bathing was infrequent. Wastes, or "nightsoil," were often collected and employed as fertilizer in the surrounding fields. Modern scientific theories of the water-borne transmission of microbial diseases had not yet been established, and people widely believed that bad odors ("miasmas" or "emanations") caused disease, but that these could be alleviated or masked through the burning of herbs or incense.[11]

This system worked well for medieval towns. But with the growth of urban areas in the West in the seventeenth and eighteenth centuries, cities became literal cesspits. The smell of human excrement began to lose its rural associations with fertility and was increasingly sensed as a symptom of disorder and decay. Amid the rapid urbanization of the Industrial Revolution, the limited availability of clean water supplies led to widespread concerns over water quality and availability. As cities grew, the belief arose that some means of supplying large amounts of water and removing large quantities of sewage was becoming increasingly necessary. In the nineteenth-century cities, universal water and sewerage networks emerged as the preferred model. Water was to be mass-produced, abstracted in large quantities, and treated at plants before being distributed through networks in densely built-up areas where economies of scale made supplying water feasible.

Modernization, in other words, implies the industrialization of water: the mobilization of large volumes of water as an economic input, a factor in public health, and a means (via sewerage networks) of disposing of the "effluents of affluence." For these reasons, as the Industrial Revolution spread around the world, so too did water supply networks. Artisanal methods of collecting and distributing water (such as pedestrian vendors delivering water jugs to individual homes) were gradually replaced by networks of pipes leading from reservoirs to consumers' water taps.[12] Sewerage systems followed, first as a vehicle for disposing of the source of "miasmas," and later as a response to emerging scientific conceptions of health and hygiene that definitively linked polluted water with diseases such as the cholera and typhoid that ravaged eighteenth- and nineteenth-century cities.

Large-scale irrigation networks were the rural counterparts to the urban water supply network. Both rural and urban supply depended on an unprecedented mobilization of raw water through large-scale hydraulic infrastructure. Initially, these hydraulic works were celebrated as technological feats, and indeed large dams and associated irrigation schemes are still mobilized as symbols in this way. But urban water supply networks—buried and thus invisible—quickly became banal, often overlooked, almost forgotten symbols of modernity.[13]

Many of these water supply networks were created by private companies. Chapter 3 expands on this discussion of the role that private companies played in water supply networks in cities. In this chapter, however, the focus is on the historical context in which water supply networks were developed, focusing on lower-income countries, for the majority of which large dams—rather than water supply—were the focus of activity during the first half of the twentieth century.

Crucibles of Development: Dams and the "Big Push"

The twentieth century was the epoch of large dams. Dams symbolized the entry of nation-states into the "club of modernity," as suggested by Nehru's eulogy to large dams as the "temples of modern India" and Nasser's proclamation that dams were Egypt's "pyramids for the living."[14] These large hydraulic works are expressions of the municipal (or in this case, state) hydraulic paradigm referred to earlier. So it is no surprise that their proliferation in the twentieth century has no historical equivalent. There were approximately forty "large" dams with reservoirs greater than 0.1 km^3 (1 billion cubic meters) at the start of the twentieth century and over 2,800 such reservoirs by the year 2000.[15] Nearly all of these dams were built with government finance, by state organizations, and were often supported by development organizations as part of the postwar "big push" for development.

This "big push" strategy was modeled on assumptions stemming from the then-dominant school of thought in development policy: namely, that "developing" countries could achieve rapid rates of economic growth by emulating the path taken by the West, following a linear pattern of growth stages. Modernization and industrialization could be effectively stimulated by selective investment in critical areas of a country's infrastructure (the big push) so as to stimulate production (leading to a "take-off" of the economy).[16] This both complemented and reinforced the reliance on state-led planning fostered by development agencies, which necessitated centralized resource management as a means of mobilizing the required inputs for industrialization and agricultural intensification.[17]

The tangible benefits of large dams were obvious: hydropower generation for electricity and revenues to the state; flood management and facilitation of navigation (particularly in rivers with highly variable flow regimes); increased supply security, as several years' supply might be impounded behind major dams; and recreation and tourism. The major intangible benefit was no less important: entry into the "club of modernity," as symbolized by Lenin's declaration in 1920 that "Communism is Soviet power plus the electrification of the whole country," as the USSR embarked on an ambitious scheme of dam building.[18] Political, economic, and broader social agendas merged to lend support to dams of unprecedented scope, such as Ghana's Volta Dam and Egypt's Aswan High Dam on the Nile in the 1960s, each of which has four times the storage capacity of the Hoover Dam reservoir (creating the two largest artificial lakes in the world).[19] Many other countries embarked on similarly extensive hydraulic works programs, usually to support irrigation for state-led agricultural development, hydropower for industrialization, or water-intensive commodity processing.[20]

Large dams enabled exponential growth in water demand over the twentieth century. Estimates suggest that at the start of the twentieth century global water abstractions (i.e., withdrawals) totaled approximately 580 km^3 per year; by the end of the century, this had increased to approximately $4,000 \text{ km}^3$ per year.[21] Population growth explains only part of this increase; the world uses more water per capita to produce food, goods, and services than ever before. Four out of every five liters of water goes to irrigate cropland. The "green revolution"[22] and the rapid growth in world food production and productivity in the twentieth century were predicated on a mobilization of water unprecedented in human history. In 1800, the world's irrigated area was 8 million hectares. Irrigated area grew to 40 million hectares by 1900, and then grew over sixfold to reach 255 million hectares (2.5 million km^2) by the end of the century,[23] by which time approximately

15 percent of world food production had become dependent on water supplied from large dams.[24] Regulating water played another critical role in modernization through the generation of electricity. Whereas waterpower generated largely from waterwheels placed in streams and rivers was the primary power supply for the initial stages of the Industrial Revolution, it was hydropower that fuelled later stages of industrialization, and which now provides approximately 20 percent of energy generation globally.

Dams were, in short, a utopian project central to modernization and industrialization efforts. The construction of dams throughout the twentieth century was impressive: by the end of the century, an estimated 60 percent of the world's rivers had been affected by dams and diversions.[25] But the negative effects of large dams became increasingly apparent over time.[26] In arguing against large dams, anti-dam campaigners asserted that their benefits—flood control, electricity generation, and increased security of water supply for industrial and agricultural users—were outweighed by the costs: extirpation of species, particularly fish; displacement of communities (an estimated 40 to 80 million people); flooding of cultural sites; contamination of water sources; disruption of ecological processes; and environmental degradation.[27] Intense international debate culminated in the creation of the World Commission on Dams (WCD), through which proponents and opponents with seemingly irreconcilable differences participated in a review process of unprecedented scope and transparency in international development.[28] The commission's findings were stark: the costs of significant negative social and environmental impacts had all too often been underestimated and had been disproportionately borne by displaced peoples, downstream communities, taxpayers, and the environment. This lack of equity, the commission stated in its report, "has called into question the value of many dams in meeting water and energy development needs when compared with the alternatives,"[29] which government officials, dam proponents, and the development community had been loath to consider.

Despite these criticisms, the allure of large dams remained high for multilateral financial agencies and governments alike. The mobilization of water resources for specific projects (from mining to irrigation), the potential of electricity generation, and the symbolic importance of large dams were powerful incentives for the big bureaucracies that financed, built, and operated them.[30] Yet the push for large dams tailed off by the 1980s because of the exhaustion of suitable sites for dam development in some countries, public protest over the high and inequitably allocated social and environmental costs, a disinclination of multilateral financial institutions to fund large-scale hydraulic works, and a more realistic appraisal of the costs and lifespan of large dams.[31] The push for big dams had passed its apogee; by the 1990s, a trend toward dam decommissioning had emerged.[32]

"Water for All": The Growth of Water Supply Lending

Large dams were thus the focus of the development sector for much of the latter half of the twentieth century. Water supply received relatively less attention. Today, this might seem surprising, as the notion that water supply should be part of the portfolios of multilateral lending and bilateral aid agencies is widely accepted. But this was not always the case; early lending activities by the World Bank and other major development banks deliberately excluded or only marginally focused on water supply.

This was in part because of the focus on large dams, consistent with the development community's early emphasis on large-scale infrastructure and relative neglect of poverty-related issues. Thus, water supply received relatively little attention within the nascent postwar development industry; this began to change only in the 1960s. During this decade, the links between water and health began to receive greater attention through the activities of the World Health Organization.[33] Interest in funding water supply projects was fueled by a new "poverty focus" among development banks. This was evidenced, for example, by the creation of the International Development Association (IDA), which offered concessional loans (termed credits) to the poorest countries under the auspices of the World Bank, signaling a broadening of the Bank's previous focus on loaning at market rates to middle- and high-income countries for postwar reconstruction via the International Bank for Reconstruction and Development (IBRD).[34]

Moreover, technical assistance was more easily available, and project identification was simplified, through increased cooperation between UN agencies on development projects, notably via the United Nations Special Fund after 1959 (and the United Nations Development Programme after 1966), but also via both formal and informal interagency agreements (such as that between the World Bank and the World Health Organization).[35] The creation of new regional development banks—notably the Inter-American Development Bank in 1959, the African Development Bank in 1964, and the Asian Development Bank in 1966—created competitors to the World Bank in markets where poverty-focused lending was the best growth opportunity.[36]

Increased investment on the part of major international lending institutions in water supply was merited, proponents argued, for several reasons.

The drive to increase agricultural productivity and nutritional intake levels, central to the developmental mission in the first postwar decades, was hindered by lack of access to safe, clean water supplies. As a result, water supply projects were frequently included within and managed under the auspices of agricultural sector loans—although this approach would later be questioned. A second reason stemmed from rapid urbanization and the associated growth of informal settlements without access to water supply (and therefore vulnerable to health epidemics). A final reason was growing concern about poverty, framed as "underdevelopment," in the context of the cold war struggle for geopolitical influence. Water supply was a legitimate concern for the new set of poverty-focused development institutions that began emerging in the late 1950s (distinct from the original emphasis on economic growth of the first wave of development institutions created immediately following the Second World War).[37]

Incentives to lend were also important in explaining the increased interest of international lenders in water. As I shall explore below, pressure from developing countries to make loans available for water supply and sanitation was a powerful inducement to extend lending activity in this area.[38] The opportunities for linking capital-intensive hydraulic works projects to lucrative contracts for domestic consultancy and construction companies were an important incentive for bilateral aid agencies, particularly those with explicit "tied aid" policies.

Consequently, water-related development lending increased, with the Inter-American Development Bank and World Bank as leading multilateral lenders. Regardless of the funding source—bilateral or multilateral, domestic or international, public or private—the majority of finance was directed through government-controlled entities. Typically, the central government developed water resources, either directly through a ministry or indirectly through a government-owned, stand-alone entity ("parastatal").[39] This approach conformed with the state hydraulic paradigm discussed in chapter 1, which dominated the water sector in both developed and developing countries and focused on subsidized, supply-side, large-scale, state-run hydraulic infrastructure projects.[40]

Water supply was, however, different from large dams in several important ways. In contrast to the centralized management of large-scale hydraulic works, water supply tended to be small scale and subject to jurisdictions at multiple orders of government (municipal, regional, and central). Business models for water utilities were also more diverse. In urban areas, public utilities were often integrated directly into government, usually as a district or municipal department or service. In both rural and urban areas, corporatized utilities or parastatals operated as public corporations, separate from local government structures. These parastatals were favored—and often created—with grants from the multilateral development banks, and particularly common in Africa (in Ghana, for example, one parastatal managed the water supply networks throughout most of the country). The World Bank's view reflected the conventional approach at the time, that "public utility projects...would do more than almost anything else to trigger development....A corollary doctrine was that autonomous authorities comparable to the Tennessee Valley Authority in the United States were needed to staff and operate the projects and insulate them from the vagaries of domestic politics."[41]

Less frequently, publicly owned companies operating under private corporate law were established, as in the Netherlands.[42] Whatever the business model adopted, public water supply was generally characterized by several common characteristics: a universal duty of supply, although in practice this often meant a focus on elite residential and industrial areas; subsidized charges (usually via cross-subsidy among water consumers or direct government subsidy to the utility's budget); and an emphasis on security of supply rather than water conservation, and on technical efficiency rather than economic efficiency. Engineering expertise dominated, and economic concerns—particularly full-cost pricing over the lifetime of expensive, albeit long-lived water supply infrastructure—were usually given only secondary consideration. The urban bias of water supply projects was based on the apparent unprofitability of rural water supply, which was relatively neglected by the multilateral financial institutions in the first decade of lending. In rural areas, projects were further complicated by competition for water resources and jurisdictional overlap between multiple sectors—notably, irrigation and hydroelectricity.

In urban areas, the municipal hydraulic approach focused on supply-side planning, under the auspices of (often centralized) government agencies, with a tendency to focus on capital works rather than ongoing operating costs and governance issues and with little attention to environmental impacts, characterized public water-supply delivery for three decades, from the early 1960s to the end of the 1980s. The UN Drinking Water Supply and Sanitation Decade (1981–90) lent impetus to the drive to increase the number of water supply projects in developing countries. Yet despite significant investments, over 1 billion people (by most estimates) remained without adequate water supplies by the start of the 1990s.[43]

Well before this, attention had begun to focus on systemic problems with the conventional model of water supply delivery. A significant number of water supply utilities were beset by a "vicious cycle": low investment, low performance, and low cost recovery limiting further investment.[44] Although comprehensive studies were few, the World Bank estimated in 1994 that only 30 percent of costs were recovered, on average, in urban water supply systems.[45] To some degree, this was because external subsidies were the norm in the sector, based on the assumption that affordability was the primary goal. There were other, less admirable reasons. As one of the largest revenue-generating agencies under the control of municipalities, water utilities in some instances became "cash cows," furnishing opportunities for cash injections into the municipal budget or for patronage through the allocation of desirable government jobs. Political incentives (which prioritized votes over raising water tariffs) were at times responsible for discouraging long-term investment as tariffs were frequently set below levels required for infrastructure renewal, and in some cases failed to cover even operation and maintenance costs.

Efficient public water utilities did, of course, exist, although they were often overlooked amid the growing rhetoric condemning government provision.[46] Moreover, most urban water supply utilities were characterized by a bias toward the urban elite,[47] emphasizing large-scale prestigious hydraulic works and distribution to wealthier urban neighborhoods, with scant attention to the needs of informal communities—whose social status, low levels of income, and location in squatter settlements without formal tenure (often on the peri-urban fringe) greatly reduced their political leverage.

Another problem was the question of fragmented jurisdiction: whereas most large utility networks were managed at the regional or national scale, many water supply utilities were managed at the municipal scale. Fragmented jurisdiction meant that local authorities could rarely combat environmental externalities (to which they also contributed), because they had limited jurisdiction over issues relating to environmental quality. Although much of this information was anecdotal (few sector-wide studies were conducted by lending agencies and bilateral aid agencies), poor performance seemed to confirm the existence of "government failure" and lent weight to arguments in favor of greater private sector involvement.

"State Failure" or "Bank Failure"?

As I have described, much of the debate in the 1980s focused on the shortcomings of water supply utilities and the governments that ran them. Less attention was paid, however, to the possible flaws in the models of multilateral financial institutions and bilateral aid agencies. Did the models

used by these organizations contribute, at least in part, to the poor performance of water supply utilities? This section explores that question in
detail, focusing on the case of the World Bank.

The World Bank is an emblematic example of the evolution of multilateral lending for water supply for three reasons. First, the Bank is the largest
of the multilateral lenders in this sector globally; in low-income countries, it
has often consistently been the largest source of MDB finance for decades.
Second, the Bank's relatively high degree of transparency to external scrutiny (in contrast to some of the other MDBs) has made a significant number
of documents—including internal evaluation reports—publicly available.
Finally, the Bank plays a leadership role in development policy in general,
and in water policy and lending in particular. Its policy initiatives frequently
set the parameters of broader debate—as they did in the case of the increasing dissatisfaction of the development community with the status quo in
water supply and sanitation at the end of the 1980s.

A full understanding of the evolution of the Bank's role must begin with
the initiation of Bank lending for water supply and sanitation, which it
undertook tentatively and only after much hesitation. Indeed, the Bank did
not engage in a significant amount of water supply lending until the 1970s.
Considering the importance attached in the contemporary period to water
supply-related development goals, this seems curious. Why would the Bank
have hesitated?

The answer lies, first, in a central challenge the Bank faced in its first decade: the need to establish its credit worthiness and obtain working capital
through bond issues. Thus constrained, the Bank's policy held that it should
not lend for "non-self-liquidating ventures"—projects with low rates of return that were unlikely to generate sufficient revenues to enable the loan to
be repaid (particularly if the investment generated revenues only in "soft"
local currency).[48]

On these grounds, it made sense to exclude water supply, sanitation,
education, and housing from the Bank's remit.[49] Moreover, the original articles of the IBRD made such lending difficult. In any event, much attention
was paid to postwar reconstruction in industrialized countries; until 1957,
more than 50 percent of World Bank financing was to developed or "Part I"
countries (although this fell to 0 percent by 1968).[50] Finally, Bank officials
were unwilling to sacrifice the accountability that they felt flowed from the
demands of generating foreign currency-denominated revenues.[51]

Accordingly, the Bank's dominant strategy focused on infrastructure
that was believed to form the basis for the development of all other sectors of the economy: irrigation and reclamation, migration, transportation,

communications, and power facilities (as well as the public utilities and training required to deliver this infrastructure). Consistent with the "big push" argument, the Bank targeted investment in infrastructure in areas "relatively less attractive to private capital" that would catalyze technological development, capital investment, and trade, and thereby result in increased production and incomes in developing countries. Poverty would thus (indirectly) be alleviated.

Accordingly, the Bank lent for big dams rather than for water supply. The Bank's first loan to a developing country was to Chile in 1947 for a hydro-electric project, and water-related hydraulic infrastructure occupied a stable proportion of the Bank's portfolio throughout the following decades. Indeed, the World Bank Group was to become the biggest cumulative lender for large dams over the course of the twentieth century, with lending peaking in the 1970s and 1980s.[52] Through its lending activities, the Bank actively promoted both centralized planning and large-scale hydraulic infrastructure. And water-related issues, particularly stemming from the Indus Water Treaty and the Aswan High Dam/Suez Canal crisis in the 1950s, did much to establish the Bank's role as a key figure in postwar international economic diplomacy.[53]

Yet several factors encouraged the Bank to change its policy. Perhaps foremost was the need to find bankable projects in developing countries (particularly as the Marshall Plan effectively sidelined the Bank's role in Europe's reconstruction, and European economies recovered). As one memo noted, "The number of water supply projects is so vast that if they were eligible for Bank financing, the increasingly difficult problem of finding suitable projects in underdeveloped countries would be greatly eased."[54] Another argument was made on the basis of the links between public health and water supply, which underpinned water supply–related activities of other international agencies, such as the World Health Organization. Proponents also argued that emphasis should be placed on "self-financing" (e.g., via user fees) rather than simply "self-liquidating" projects; lending for the former types of projects would allow the government to direct funds to sectors for which self-financing was not considered to be possible or desirable (like basic education).

Despite these arguments, certain Bank officials were reluctant to lend for water projects on the basis of their overriding commitment to economic growth per capita, which might be compromised by better health outcomes arising from improved water supply. As one internal memo noted:

> More attention should be given to the possible unfavorable side effects
> of improved water supply. The usual studies tend to emphasize better

health, longer life, and greater productivity management. The negative effects of the resulting higher rates of population growth, and of changes in the age structure, on available income per head or on overcrowding, are seldom brought out. Insistence on a more balanced treatment would help in placing water supply projects in better perspective.[55]

From this perspective (outrageous from a contemporary standpoint), water supply investment should not be pursued, insofar as consequent reductions in mortality would reduce income per capita, and hence reduce quality-of-life indicators (on the dubious assumption that total GDP would not increase if population health increased).[56] This (albeit extreme example) hints at the debates taking place within the Bank, and the development community more broadly, regarding water supply lending.

Yet by the late 1960s, other arguments had won the day. Using language that would have been "dismissed as unpersuasive and unacceptable a few years ago,"[57] the World Bank's 1971 annual report stated that

> a major reason why the Bank Group has tried to encourage the development of this [i.e., the water and sanitation] sector is that its benefits are so pervasive. It is generally accepted that the financial returns accruing to water and sewerage authorities considerably understate the true benefits to the community, by no means all of which are readily apparent or easily quantifiable. Environmental improvement, a valuable end in itself, can also result in additional financial benefits by improving the community's productive capacity through raising or safeguarding the level of public health.[58]

The creation of the International Development Association (IDA) in 1960, with its focus on "social lending" and "soft loans" to sectors such as agriculture, housing, water supply, and education, enabled the Bank to pursue this agenda. This paralleled a broadening of focus on "poverty alleviation" within the Bank, which implied a much more heterodox approach to lending as compared with its earlier emphasis on increased productive capacity (economic growth, narrowly defined) and its consequent emphasis on large-scale infrastructure.[59] By the late 1960s, the World Bank was "the leading proponent of the view that investment in...public utilities was a precondition for the development of the rest of the economy."[60] This view was consistent with the Bank's broader understanding that an adequate supply of public infrastructure (railways, roads, power plants) was a necessary condition for the stimulation of private investment.

A corollary policy was that autonomous government bodies, analogous to the Tennessee Valley Authority whose model the Bank had frequently touted, were needed to run these projects, in order to "insulate them from the vagaries of domestic politics."[61] Indeed, the Bank actively supported the creation of a significant number of autonomous entities in developing countries, including those responsible for hydraulic works. These agencies served several purposes for the Bank: operating outside the normal avenues of legislative and judicial oversight, staffed by those sympathetic to the Bank's policies, these agencies "created a steady, reliable source of what the Bank needed most—bankable loan proposals—as well as critical power bases within nations to transform and influence the direction of their economic and social development."[62]

The Early Years of World Bank Water Supply Lending

The Bank thus began lending for water supply projects in the mid-1960s, albeit initially at a slow pace, with only two projects initiated between 1961 and 1970.[63] A significant constraint was the early focus on revenue generation rather than broader public-health concerns.[64] Contrary to expectations, most water supply lending during the decade originated from the IBRD rather than the IDA, given the emphasis of Bank staff on sufficiently high rates of return and revenue generation. Throughout this period, only urban water supply projects were funded, excluding rural areas despite their acknowledged need. Indeed, in response to internal critics of the "social" character of water supply lending, most urban water supply projects were selected based on their ability to recover costs through charging for services (in principle if not in practice). The need to prove within the Bank that these projects were "bankable" meant, however, that poverty alleviation received little attention of any substance.[65]

As a result, water supply lending tended to avoid the poorest communities, through explicitly prioritizing urban projects, and by requiring projects to have relatively high, stable revenue flows (more feasible in urban areas because of economies of scale).[66] This meant that large segments of the population—notably peri-urban and rural—were often excluded from water supply projects. The first-ever review of the water sector by the Bank in 1971 presciently warned that efficiency and financial and economic sustainability would be difficult to achieve in this sector, in part because lack of ability to pay implied either loss-making projects or the exclusion of large segments of the population.[67] This cautionary message went unheard, however, in the following decade.

With the arrival of Robert McNamara as president of the Bank in 1968, a new focus on poverty alleviation and associated expansion of the Bank's mandate and activities led to an increase in activity in the water supply portfolio. More was lent for water supply in 1971 than in the entire previous decade,[68] and water supply lending jumped from an average $27 million per year during the period 1968–70 to $180 million during the period 1971–73.[69] One reason for this rapid increase was that water supply projects were sufficiently comparable to traditional economic infrastructure that they seemed to satisfy both the new "social" and traditional lending criteria. Moreover, the Bank's formal agreement with the World Health Organization in 1971 on water supply and sewage disposal (in which the WHO cooperated closely with the Bank in identifying suitable projects for funding) facilitated its drive to increase lending in this sector, as did the creation of the UNDP-WB Water and Sanitation Program in 1977, following the United Nations Water Conference at Mar del Plata.[70]

The Bank's involvement in the water sector increased and broadened to rural areas in the 1980s, encouraged in part by the goals set during the UN Drinking Water Supply and Sanitation Decade, and in line with the overall growth in so-called poverty lending by the Bank. By the 1980s, water loans were averaging 4 percent of the Bank's portfolio. Investment in the water and sanitation sector grew in importance: by 1988, the Bank's water projects represented 10 percent of total projects.[71] The first loan entirely dedicated to rural water supply was signed in 1977.[72]

Yet growing concern had been voiced within the Bank regarding the sustainability of water supply projects. Bank staff emphasized the poor performance of water supply and sanitation projects relative to the average in the Bank's portfolio.[73] The development community tended to focus on the need for greater community participation in water supply management as a means of improving equity of distribution of costs and benefits, cost recovery, monitoring, and maintenance of facilities.[74] NGOs tended to emphasize the lack of transparency and accountability in water sector projects—with negative effects on project sustainability and on equity, particularly where displacement and resettlement occurred in the case of large dams. This critique was often situated within a broader set of NGO campaigns for greater attention to environmental issues, increased community participation, and greater accountability and transparency by the Bank, notably in light of the results of the damning independent review—commissioned by the Bank—of the Bank's involvement in the controversial Indian Sardar Sarovar (Narmada) megaproject.[75]

As a result, the Bank commissioned a review of its experience with water supply and sanitation lending, covering 129 urban water supply and sanitation projects over the period 1967 through 1989.[76] Released in 1992, the "Buky report" was a scathing indictment of the Bank's performance. Its findings echoed those of the "Wapenhans report," released in the same year, which was highly critical of the Bank's approach to project lending, and pointed to particularly poor performance in the water sector.[77] Numerous internal reports had, in fact, documented the decline in portfolio quality throughout the 1970s and 1980s, due to a set of interrelated factors: failure to enforce key loan covenants (such as debt-service coverage and rates of return), the inability of generalist staff at the Bank to monitor and enforce good practice, the lack of technical assistance, and an "approval culture" linked to the pressure to lend.[78] The result was a "major embarrassment" in sectors in which the Bank had built its earlier reputation: infrastructure lending for power, transport, major irrigation systems, water and sewerage, and telecommunications.[79] The water sector was one of the least satisfactory of any of the Bank's portfolios, with some of the lowest sustainability ratings of any sector, as assessed by the Bank's own operations and evaluation department.[80]

The Buky Report: An Analysis of Failure

The Buky report dissected the causes and results of these failures in devastating detail. The Bank, it stated, had failed to achieve success in any of its four principal objectives for the water supply and sanitation sector. It had "rarely succeeded" at institution building; failed to help institutions achieve financial viability (with a 78 percent failure rate in respecting financial covenants on loans); failed to ensure the availability of a minimum supply of safe water to the poor (only two out of 129 projects demonstrated success); and delivered water supply and sanitation infrastructure but rarely ensured long-term sustainability through effective operation and maintenance programs.[81]

The symptoms of failure were numerous. For example, with only a few exceptions, "unaccounted-for water" (UFW) (the difference between the amount of water put into the system and that which yields revenues—due to leaks, incomplete billing, or corruption) increased rather than decreased in the majority of projects. One review found that the average volume of UFW six years after project inception was 20 percent higher than forecast, leading the report to conclude that an "unacceptably high amount of non-revenue water is one of the main characteristics of the water supply projects financed by the Bank and IDA."[82] The Manila Water Supply and Sewerage

Project was an extreme example: with UFW of 40 to 42 percent at the start of project and about 66 percent at end of project, "the entire output of this project was [lost to] unaccounted-for water."[83] Lack of attention to environmental issues was another symptom of failure; few project design and planning documents, for example, even mentioned possible environmental and public-health impacts; this absence was surprising, the report suggested, given that most projects intended, but were ultimately unable, to include a sewage treatment component.

The lack of technical capacity and incentives on the part of World Bank staff was identified as a proximate cause for these failures. The underlying causes were more systemic: pressure to lend, leading to a lack of attention to factors critical to long-term sustainability; lack of managerial autonomy and capacity on the part of borrowers; absence of Bank policies requiring loans to meet environment and poverty-related criteria; unwillingness by the Bank to impose and act on enforcement mechanisms (e.g., loan covenants); and a short-term, narrow approach on the part of the Bank and borrowers, focusing on individual projects rather than a long-term, integrated approach to the sector as a whole.[84]

The Buky report's criticisms echoed NGO critiques of the Bank, including its bias toward large-scale centralized development projects and the negative impacts of its projects on disadvantaged groups and the environment.[85] Although the report's recommendations diverged from the then-radical alternatives proposed by NGOs (prioritizing demand management and water conservation over new projects; focusing on sewage treatment and pollution control; supporting small-scale projects such as rain-fed agriculture rather than megaprojects), some NGO recommendations were included—notably, an integrated sectoral approach and more extensive monitoring of performance with respect to impacts on disadvantaged populations and the environment.

The Bank's Policy Shift and the PSD Strategy

The resulting shift in policy was captured in the World Bank's new strategy on water resources, released in 1993.[86] Its call for integrated water resources development and the prioritization of broader sectoral reform signaled a move away from previous policies. Specifically, the new strategy confirmed the Bank's shift away from support for parastatals, which had begun with the Berg report of 1981, and was consolidated over the 1980s with the Bank calling first for the reform and then elimination of parastatals.[87] Notably, decentralization and privatization were key elements of the reform agenda. This was consistent with the Bank's emerging emphasis on enabling private sector development and greater use of market mechanisms, as symbolized by the Bank's adoption of the Dublin Principles (including, controversially, the assertion that water should be recognized as an economic good) in 1994.[88]

The weakness of the Bank's water supply portfolio, as explored below, made it ripe for realignment with the Bank's emerging private sector development (PSD) strategy.[89] The PSD strategy signaled an abandonment of the earlier McNamara doctrine that "strengthening the market and quality of management, not public versus private ownership, was the dominant industrial policy consideration for the Bank."[90] The strategy resulted, in part, from the Bank's repeated experience of "failed" attempts to reform state-owned enterprises; new economic theories of "state failure" and techniques for introducing competition in sectors previously understood to be natural monopolies; and the "demonstration effect" of successful restructuring and privatization in OECD countries, notably the United Kingdom, the United States, and New Zealand.[91] External pressure, notably from the United States, created an additional imperative for the Bank's increasing emphasis on private sector development.[92]

Although eager to increase private sector–related activity, the IBRD and IDA were prohibited from lending to nonsovereigns (organizations other than sovereign governments) by a clause in the Bank's original Articles of Agreement.[93] Nonetheless, the World Bank Group began to develop other activities more directly supportive of the private sector. The provision of investment guarantees through its Multilateral Investment Guarantee Agency (created in 1988), together with the dispute arbitration and conciliation services offered through the International Centre for Settlement of Investment Disputes (created in 1966, but little used until the late 1980s), were intended to increase investor confidence and thereby facilitate foreign direct investment in developing countries. Direct investment in private sector water-supply projects on the part of the International Finance Corporation (the part of the World Bank Group of affiliates mandated to invest directly and solely in the private sector)[94] increased dramatically in the mid-1990s, while concessional IDA loans for water and sanitation decreased (figure 2.1).

By the mid-1990s, the pendulum had swung within the Bank from the McNamara doctrine to a more explicit preference for private sector involvement. The water sector followed suit. The "municipal hydraulic" paradigm was to be replaced by the key tenets of the "market environmentalist" paradigm (explored in chapter 1): treatment of water as an economic good; cost recovery and full cost pricing-based tariffs; elimination of cross-subsidies where

2.1. *World Bank lending for water: IFC, IBRD, and IDA commitments, 1964–2006 (2006 US$) (Bakker et al. 2008; data drawn from World Bank Group annual reports)*

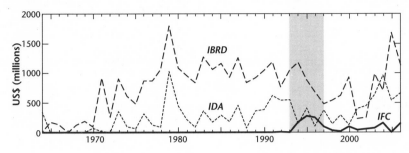

possible; water rights (and water markets, where feasible); decentralization; commercialization and/or privatization; and a change in the role of the state from operator to regulator, with a key role in standard setting.[95] These reforms—changes in water-related legislation, reforms of business models for water supply utilities, introduction of cost-related tariffs for irrigation and drinking water—were encouraged through a new approach to water supply: sectoral lending, in which sector reform was required before additional loans could proceed. In some cases, such as that of Indonesia, even structural adjustment loans were predicated on water sector reform.

However, the core challenges posed by water supply—the difficulty of balancing revenue generation with access, lack of ability-to-pay and public-health externalities associated with water supply, and high capital intensity combined with low economic rates of return—were not addressed by the Bank's new strategy. The Bank's own assessments had revealed that its conventional approach to water supply (capital-intensive foreign currency loans to public utility/parastatal structures, focused on capital works) had not succeeded in addressing these challenges. This omission, as we shall see in chapter 3, came back to haunt the Bank at the end of the 1990s, when substituting private ownership and management for management by the public sector did little to resolve these core challenges.

THE EMERGENCE OF THE PRIVATIZATION AGENDA

The evolution in the World Bank's policies was emblematic of, and played a catalytic role in, the shifts in lending and policy for water supply in developing countries throughout the 1990s. Although the Bank had acknowledged the important role of the private sector throughout its history, in practice its activities were almost exclusively focused on government activity.[96] During the 1980s, the Bank expressed greater support for pro–private sector policies, followed by a more active private sector development agenda in the 1990s, triggered by its private sector development strategy.[97] Other multilateral development banks followed suit, as did bilateral aid agencies, several of which had initiated pro–private sector initiatives that predated those of the Bank (notably, the United States Agency for International Development and the UK's Department for International Development).[98]

Support for privatization must be understood in the context of a new set of policy priorities adopted by many multilateral financial institutions in the 1990s. This set of policies—colloquially referred to as the "Washington Consensus"—included financial liberalization, privatization, deregulation, the creation of secure property rights, tax reform, the introduction of competition, and public sector fiscal "discipline" (including the selective redirection of public funds toward "public goods" that facilitate income redistribution, such as education, primary health care, and infrastructure).[99] This so-called (and short-lived) consensus emerged as a response to the fiscal crises experienced by the governments of many lower- and middle-income income countries in the 1970s and 1980s, characterized by high indebtedness, high inflation rates, and slow or negative growth. These fiscal crises provided leverage that multilateral donors (most importantly, the World Bank and IFC) could use to impose reforms.[100] At the same time, the increasing integration and deregulation of global financial markets made possible private financing of increasing scope and scale.[101] This lent greater credibility to the ideological arguments pertaining to "market failure," particularly in the context of the end of the cold war (and the shift in aid flows implied by the resulting realignment of geopolitical allegiances). It also added credibility to the shift from lending to governments to channeling finance via the private sector. The European Bank for Reconstruction and Development (established in 1990) is mandated by its charter, for example, to carry 60 percent of its portfolio as direct loans to the private sector.[102]

A key driver for privatization thus seems to have been the negative experience of the IFIs with lending to the water supply sector in the 1970s and 1980s, combined with a more general shift of ideological direction and strategy. From this perspective, involving private firms made sense, as they could improve performance and increase efficiency. The discursive justification for the need for improved efficiency was multifold, and subtle: efficiency is a necessary strategy because water is an increasingly scarce resource, requiring the efficient allocation that only the market can provide. Water scarcity, from this perspective, is a universal condition that is at once both natural (which justifies cost-reflective pricing and the commercialization of water) and social (which is the result of flawed public management and thus justifies private sector management and ownership). But policymakers largely overlooked the fact that efficiency would arise not only through the more flexible financial engineering and greater expertise to which the private sector was presumed to have access, but also through removing social policy goals (such as employment generation or income redistribution through cross-subsidy) from water policy. Yet this stance, particularly its myopia concerning the effects on poor households, was to become a major source of critique of private sector involvement. Another criticism was directed at the shift from defining water as a public good (which is subject to market failures and must be supplied by the state at subsidized prices) to an economic good (which can profitably be supplied by the market under competitive conditions). As explored in chapter 1, water has economic value, but its treatment as an economic good is complicated by externalities (public health and the environment) and quality issues.

In the 1990s, however, these critiques had little impact on the major development banks. The provision of much-needed capital, the introduction of competition for the market as a spur to efficiency, and the overriding need to introduce full-cost pricing were all justifications for private sector finance and management.

Another important trend was the decline in official development assistance (ODA) for water supply and sanitation in the 1990s, after three decades of steady increases (figure 2.2).[103] This was in part associated with the overall decline in total ODA from 1991 through 1997,[104] but also reflected a change in sectoral priorities: although multilateral aid increased slightly after 1998, bilateral aid for water supply and sanitation dropped sharply.[105] This decrease in aid flows for water and sanitation occurred despite the identification of water supply and sanitation as a high priority for the development community. But proponents hoped that private sector financing would, as the World Bank noted in its *Global Development Finance 2000* report, bridge the gap: "With declining foreign aid, privatization and foreign investment will help heavily indebted countries that can no longer sustain loss-making enterprises and need revenues to fund their budgets and current account deficits."[106]

These factors led to a major change in direction on the part of international financial institutions. Water supply, like other networked utilities,

2.2. *Water and sanitation: ODA, DAC commitments, 1982–2003 (data drawn from OECD Development Assistance Committee, International Development Statistics online, http://www.oecd.org/dac/stats)*

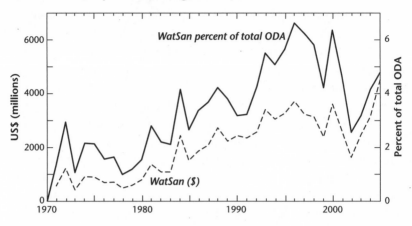

was no longer to be funded by loans or grants (on the assumption that this would free up capital for other "nonliquidating" investments such as primary education); rather, water supply utilities were to be privatized through divestiture (where possible), or managed by the private sector where divestiture was not possible. The high levels of indebtedness of many of the water parastatals were a proximate cause of privatization; bringing in the private sector would, lenders hoped, improve loan repayment rates and—importantly in light of declining net disbursements by the Bank—allow financially healthier water utilities to continue to borrow. Although critics challenged the forecasts of large borrowing needs—based on a presumption of the use of conventional, capital-intensive rather than alternative, low-cost technologies—the need for new capital sources remained an important justification, in many instances, for involving the private sector.

The desire for a transition from public to private finance and from public to private operators, ongoing throughout the 1990s, occurred against the backdrop of debates within development economics on the appropriate role of governments in economic development. The controversy over the causes of the "East Asian miracle," which proponents of Washington-Consensus orthodoxy portrayed as an exemplar of their theories, was the most visible aspect of this debate. Critics pointed to evidence of the strategic role of governments in actively intervening in the market with the goal of fostering economic development in order to argue against central tenets of the Washington Consensus.[107] Key figures like Joseph Stiglitz (chief econo-mist at the World Bank from 1997 to 2000) advocated a "post–Washington Consensus,"[108] arguing that cost and information externalities resulted in important market failures—via poorly functioning institutions and absent or incorrect incentives—that required government intervention.[109] Others challenged the export to developing countries of Western models of public administration, including the "new public management," the conceptual basis for recommendations to commercialize and/or privatize, arguing that these models were inappropriate and, if applied, could result in unexpect-edly negative outcomes.

In the water supply sector, these theoretical critiques were reinforced by the nascent antiprivatization campaigns launched by NGOs in the late 1990s (discussed in greater detail in part 2 of this book), which began to have an impact on development-community policymaking in the 2000s. Yet in the 1990s, this debate had little immediate impact on development practices on the ground. Privatization continued to be actively advocated by bilateral aid agencies.[110] The absence of concessional aid funding gave governments little choice. Moreover, extremely active market development by private water companies was facilitated by a range of training, coordina-tion, and networking opportunities created by lending agencies. Examples of such agencies were the World Bank's Rapid Response Unit (and, later, its Public-Private Infrastructure Advisory Facility); and new water-focused NGOs with a promarket orientation such as Business Partners for Develop-ment (cosponsored by private water companies and the World Bank), the World Water Council (initiated by French private water companies), and the Global Water Partnership (originally chaired by a WB vice president and sponsor of the influential interministerial World Water Forums initiated in 1997). The victory of "market triumphalism" seemed assured; yet, as we shall discover in chapter 3, the eventual gains to be made from water privatization were much more ephemeral than it seemed in the mid-1990s. By the early 2000s, the World Bank was admitting that its earlier strat-egy had met with setbacks, although it remained committed to its agenda favoring a reduced role for government and an expanded role for markets, private companies, and market-simulating regulatory and management techniques.[111] And the rise to dominance within water policy circles of the World Water Forum meetings (organized outside the United Nations system by a private nongovernmental organization with close links to the World Bank) lent further weight to the charges of critics that international water policymaking had also been privatized.

CONCLUSIONS: THE ORIGINS OF WATER SUPPLY PRIVATIZATION

In this chapter, I have argued that the growing trend of water supply privatization in the 1990s must be understood with reference to the evolution in water policy and practice over the course of the twentieth century. The shortcomings of what I have termed the "municipal hydraulic" paradigm are key to understanding this evolution. To recap, the municipal hydraulic paradigm advanced by development agencies often produced highly inequitable, indebted, and poorly functioning water supply systems. The reasons for this were rooted partly in the biases inherent in development aid toward ambitiously large-scale, capital-intensive, centralized hydraulic projects. This focus led to a neglect of issues of distribution, management, and equity.[112] Another reason for poor performance was the fact that water (and particularly water supply) was a poor fit with the conventional twentieth-century model of development. Water supply loans were non-self-liquidating, often did not generate hard currency (exposing borrowers to currency risk), generated relatively little revenue in comparison to their capital costs, and were especially subject to what would now be termed "governance failures," given the fact that they were subject to particularly fragmented jurisdiction (multiple uses and multiple scales). It is thus of little surprise, in retrospect, that many water supply projects mediated through the development industry should have performed poorly.

This perspective is not, however, intended to lay all of the blame at the feet of lenders, donors, or governments. Rather, the intent is to provide a corrective to standard government-failure arguments. This echoes the concerns of recent studies of governments and governance (raised in chapter 1), which focus on decision-making processes, interactions between government officials and recipients of government aid, and the strategies via which coalitions within and beyond the public sector shape public policy.[113]

In short, rather than focusing solely on the failings of government, we should examine the flaws in the particular model of development that the development community, as a whole, favored and promoted. In other words, I have argued in favor of a hypothesis that "state failure" was conjoined with, and in part an outcome of, a specific type of development paradigm. But this hypothesis received little attention in the debate in the 1990s, as ideology conveniently converged with pragmatic justifications: the need of lenders to recoup investments and resuscitate insolvent water-supply utilities, the emerging supply of private sector expertise fostered by liberalization of the water sector within OECD countries, the drive of

European private water companies to internationalize in the context of se-cure but low-profitability domestic markets, together with the increased flows of "foreign direct investment" to developing countries in the 1990s that seemingly obviated the need for concessional development finance, particularly to middle-income countries.

All of this increased the vulnerability of governments to the demands of donors and lenders for privatization in the 1990s. Why, then, did privatiza-tion, and market environmentalism more generally, emerge as the solution? On the one hand, the drivers were pragmatic. In some cases, water supply utilities were too indebted and poorly performing to borrow any further; private sector involvement provided a means of continuing to repay loans but also of creating the conditions for multilateral development banks to continue to lend to the sector. On the other hand lay clearly ideological drivers: "state failure" and the neoliberal ideology with which it was as-sociated. The drivers for privatization, in other words, were a mixture of pragmatism and ideology.

What does this argument contribute to the debate? First, we can dismiss the claim that privatization occurs for simply ideological reasons. There were (and are) significant, real problems with water supply around the world. A response was merited. Second, we can query the nature of the response that dominated development policy in the 1990s. Specifically, resuscitating the alternative hypothesis—that both state failure and market failure were at work, and that water supply subverts conventional development models predicated on one or the other of these concepts—will allow me, in subse-quent chapters (and particularly in chapter 4), to return to the question of whether we can transcend the public-private debate in the search for new models of public goods provision. Meanwhile, in chapter 3, I turn to an analysis of the rise and partial retreat of the private sector from water sup-ply in developing countries over the past decade.

3

WATERING THE THIRSTY POOR

THE WATER PRIVATIZATION DEBATE

June 1, 2005

It was an otherwise unremarkable morning in City Water's offices on the outskirts of Dar es Salaam, Tanzania's largest city. Executives at the private company—a subsidiary of Biwater, a British water services multinational—were well into their daily routine. No one was expecting the police, let alone the subsequent deportation of three British expatriates managing the city's water company.

Admittedly, Biwater had recently been the subject of harsh government criticism. For the past two years, it had been managing the city's water supply system under the terms of a ten-year lease contract, signed in 2003 by the government of Tanzania as a condition of receiving loans from international financial institutions.[1] Proponents had promised that involving the private sector would solve Dar es Salaam's water supply crisis. For Biwater, the opportunities for profit seemed promising: the city's dilapidated water-supply network had serious quality problems and had not kept pace with the city's growth, failing to reach two-thirds of the city's 3 million people. But the promised capital and infrastructure had failed to materialize, even while water prices rose dramatically following Biwater's arrival.

Biwater's poor performance quickly became the focus of high-profile public debate. The international aid community had almost universally supported the "private sector participation" contract, after successive governments had proven incapable of extending the water supply network despite decades of external aid. Involving a private company in managing Dar es Salaam's water supply system seemed like a perfect opportunity to "do good while doing business." After all, hadn't cities like Buenos Aires, London, Jakarta, and Manila taken the plunge and invited private multinationals to manage their water? Adding to pressure from international financial institutions, some bilateral aid agencies had promoted the deal; a pop song (backed by the British government's Department for International Development, a keen proponent of private management of water supply systems in Africa) proclaimed: "Young plants need rain, businesses need investment. Our old industries are like dry crops and privatization brings the rain."[2]

Despite this optimistic rhetoric, Biwater's performance had led to tense negotiations between the company and the Tanzanian government, which had recently stalled. To make matters worse, local campaigners had linked up with international antiprivatization groups as far-flung as Canada and Uruguay and were calling for the contract to be canceled.[3] Public support for cancellation was high, particularly as water had been free in Dar es Salaam until 1991, and the involvement of Biwater was viewed as part of a broader commercialization of water supply rejected by social movements. Back in the United Kingdom, media reports critical of both Biwater and the British government had been circulating, sparked by lobbying from the World Development Movement (WDM), one of the UK's most prominent development-industry watchdogs. In mid-May, the Tanzanian government issued a cancellation notice for Biwater's contract, the legality of which was disputed by company representatives.

Buoyed by public discontent, Edward Lowassa, the Tanzanian water minister at the time (and soon to become prime minister), chose a dramatic method of forcing Biwater from the country. The announcement of the cancellation of the contract at a televised press conference was followed by the surprise police visit and the detainment of City Water's three senior British expatriate managers. After being held for several hours, the three men were escorted to Julius Nyerere International Airport and placed on a plane bound for London.

Biwater's response was swift. Within weeks, the privately held company—owned by Adrian White, a multimillionaire ex–BBC governor, who had become wealthy providing services to Britain's water industry following its privatization by Margaret Thatcher in 1989—had placed advertisements in several prominent African publications accusing the Tanzanian government of corruption. "When aid flows through political pipes," the advertisement said, "it sometimes leaks." Biwater then announced that it would sue the government of Tanzania—one of East Africa's poorest countries—for between $20 and $25 million for "actions amounting to expropriation of its investment, assets and revenues in Dar es Salaam"—equivalent to two years' worth of revenues for the city's water supplier. "The Government of Tanzania has stolen our assets, broken the terms of its contract and acted disgracefully towards our staff," proclaimed a company press release.[4]

The case was to be heard by a little-known affiliate of the World Bank, the International Centre for Settlement of Investment Disputes (ICSID).[5] A decade or two earlier, it would have been difficult for a private company to *sue a government through an international tribunal. But the proliferation since the mid-1980s of bilateral trade treaties designed to facilitate foreign direct investment and international activities by private companies had created mechanisms for legal proceedings of this kind.[6] Critics argued that ICSID's secret proceedings represented an unacceptable "privatization of justice," emblematic of "globalization from above," carried out in the name of private capital. But a bilateral trade treaty the Tanzanian government had signed with the UK government, as part of a broader package of trade liberalization encouraged by its donors and lenders, obliged it to submit to ICSID's jurisdiction. In July 2008, ICSID delivered its ruling: the Tanzanian government had indeed violated its trade treaty with the UK by expelling City Water, but no damages were to be awarded because the company's value was "nil" at the time of expropriation.[7] Meanwhile, a separate legal case had been brought against City Water by the Dar es Salaam Water and Sewerage Authority; heard under the auspices of the UN Commission on International Trade Law (UNCITRAL), this ruling also went against the private company.[8]*

The victory won by the antiprivatization campaigners in the court of public opinion had been confirmed a few months earlier, when the British government's secretary for international development, Hilary Benn, announced that the department would now support "public-private partnerships" (PPPs) for water as opposed to the "private sector partnerships" (PSPs)—many of them involving British water multinationals—that it had previously promoted around the world. The promise of privatization, it seemed, had ebbed.

But for the residents of Dar es Salaam, the departure of Biwater was no victory. Their dreams of accessing water supply remained as distant as ever. Less than a third of households—mostly in wealthy neighborhoods—have access to the water supply network. Most families purchase water from those with network connections or private wells. Decades of government promises to extend water supply to poor communities have led only to disappointment. The government-run water-supply network is technically bankrupt, without funds to maintain the existing system, much less to connect poor families.

As a result, a host of informal solutions have sprung up, from private water vending to illegal connections surreptitiously connected to government pipes by so-called midnight plumbers. The result is a spaghetti-like tangle of exposed plastic pipes and tubing snaking through city streets and alleys, delivering water of dubious quality and reliability. The draw on the main network of this "structural leakage" (as policy documents coyly term it) is so substantial that networked water customers often receive only a trickle of water through the pipes. And water in the pipes is often polluted from the inevitable infiltration that results from low pressure. So customers with network water connections are forced to seek alternative methods of water supply and turn to "spaghetti connections" or booster pumps that further reduce network pressure.[9] Wealthy residents buy bottled water or drill their private wells ever deeper, but these are options only available to the few who can pay.

These individual, informal, and often artisanal modes of water provisioning are the lifeline of water supply access in Dar es Salaam and are typical of the resourcefulness of those without access to government services. Yet, ironically, these "spaghetti networks" undermine the industrialized piped water network intended to provide a collective response to the city's water supply needs.

The situation in Dar es Salaam is typical of cities around the world. Despite decades of effort, the government has not achieved universal water supply, or anything close to it. But the private company brought in to manage the network did not provide a sustainable alternative. The majority of urban residents are left to their own devices. In an article documenting similarly difficult conditions in Africa's largest city, Lagos, British journalist John Vidal quotes a local resident's despairing comment: "People wake up angry in a society like this."[10]

As the Biwater case suggests, water privatization inspires fierce controversy. Those opposed to privatization argue that profit-driven management will not improve water supply access; that it is unethical to make a profit supplying people with a resource essential for life, ecological health, and human dignity; and that private management will create environmental harm (through, for example, worsening water scarcity and pollution).[11] Proponents of privatization argue, in contrast, that introducing private management of urban water supply networks is an urgently needed strategy of "revolutionary change," which is our best option for supplying water to the world's urban poor.[12] Commercializing water, they contend, will encourage the conservation of scarce water resources through higher prices, thereby discouraging wasteful overuse.

This debate typifies broader disagreements over the state-market relationship and the "marketization of nature." As we have remarked, this debate pits partisans of classic forms of state intervention against "neoliberals," whose reformulation of the role of the state emphasizes the need for selective regulation but not government provision of public services. It

also pits proponents of commercialized forms of environmental management against those who argue for alternative worldviews of water use and
protection.

Part 2 focuses on these alternative worldviews. To set the stage for this
discussion, this chapter explores the debate over water privatization in
urban areas in developing countries. This focus is merited for two reasons.
First, cities in developing countries have been at the frontier of privatization, as sites of a rapid expansion of private water management over the
past two decades. Second, the most acute expression of the world's water
supply problems is now in urban areas: severely degraded water resources,
highly inequitable access, and a scarcity of fresh, safe, reliable water.

It is important to note that this is not the first time that many cities have
experienced private management of their water supply network. In order to
provide some historical perspective on the debate, this chapter first briefly
revisits the nineteenth-century experience with private sector ownership
and operation of urban water networks, which ended with the municipalization of most water supply systems around the world.

The resurgence of private sector involvement since the 1990s is the
subject of the remainder of the chapter, which focuses on the question of
whether private companies can realistically be expected to supply water
to the "thirsty" unserved poor. The analysis presented here examines evidence for the two main claims made in favor of privatization.[13] On the one
hand, proponents advance a fiscal argument: governments will be relieved
of the burden of public debt when the private sector is mobilized to provide
financing. On the other hand, proponents make an efficiency argument,
based on economic theory (most importantly, assumptions deriving from
"public choice" about the behavior of government managers): the performance of water utilities under private management should improve. This is
often associated with a "pro-poor" argument: more financing and greater
efficiency should enable more of the (urban) poor to access water supply.
In this chapter, I examine reasons for the failure of the private sector to
achieve these goals and document the strategic retreat of the private sector
from water supply to the urban poor.

My analysis continues to suggest that conventional definitions of water
as "public" or "private" merit scrutiny. Essential for life, nonsubstitutable,
and characterized by significant public-health-related externalities, water
supply is often portrayed as a paradigmatic public good. Yet on closer in-
spection, its status is more ambiguous. To be sure, water supply has many
aspects of an economic good. Indeed, the view of water as an economic
good has gained increasing currency in international development circles

over the past two decades. Yet competing views of water—based on spiritual, ethical, or environmental criteria—continue to be voiced in opposition to this notion. However, the term "public" is, I will argue, an inadequate means of capturing the ambiguous status of water, which is simultaneously an economic and public good, an industrial input and an ecological necessity; a religious symbol and, at least in part, a commodity. This ambiguity is one of the central reasons why, as I argue below, the universal privatization of the world's urban water supply systems (as envisioned by proponents) has not and indeed never will be achieved.

THE NINETEETH-CENTURY EXPERIMENT: PRIVATE SECTOR PROVISION OF WATER SUPPLY

What can the historical experience with water privatization teach us?[14] As rapid urbanization took place in the nineteenth century (particularly in North American and European cities), urban water supply networks grew quickly. The benefits of household water connections for public health and hydrants for firefighting capability were important drivers of network development; property-insurance premiums, for example, were much lower in U.S. cities that had established water supply systems.[15] Some networks were managed directly by municipal governments; others were built and maintained by private companies. Private companies operated in Boston, New York, London, Paris, and Seville, typically supplying water to wealthier neighborhoods. A similar pattern of mixed public and private activity was evident in colonial cities, from Delhi to Jakarta, Lagos to Buenos Aires. For the poor, it was more a matter of using water vendors, public taps, wells, and rivers; or in the most desperate cases, stealing the water they needed for survival.

The tendency of private companies to concentrate in wealthier neighborhoods was a concern voiced by nineteenth-century reformers. The need to extend water supply to all citizens and neighborhoods in order to protect against the fires and water-borne diseases that periodically raged through nineteenth-century cities was a strong argument in favor of government provision and was championed by public-health reformers and municipal socialists—most notably the "gas and water socialists" in Britain and the United States. In most places, by the mid-twentieth century, governments had become owners (and, usually, managers) of water supply systems, and they extended water supply network access to all citizens. In places where private companies continued to operate—as they did most notably in France and Britain—they were tightly regulated.[16] Private water companies in the UK, for example, had dividends capped and were required to reinvest any remaining profits in the water supply business.[17]

The nineteenth-century case illustrates the difficulty of combining private sector ownership and management with the control of water supply–related "externalities." The terrible fires and water-borne disease epidemics (notably cholera and typhoid) that regularly swept through nineteenth-century cities provided a strong justification for universal provision of potable water supplies at sufficiently high pressures.[18] But the high prices charged by private companies meant that subscription to water supply networks was selective rather than universal; for example, in the 1840s in England, only 10 percent of the population had a piped water supply.[19] As unregulated monopolies, companies typically charged high prices and undersupplied water to maintain the price.[20] Moreover, sewerage systems were so capital intensive, and of such low profitability, that private water companies did not, as a rule, build them. As a result, adequate sanitation was almost completely absent in most cities, with well-documented public-health implications.

Municipalization of water supply services in the second half of the nineteenth century stemmed in part from concern over these issues. This was driven in many cases by new legislation placing public-health responsibilities with municipal governments, combined with an apparent inability or lack of interest on the part of the private sector to finance universal provision (in other words, network extension to all customers, of all incomes). The standards of service being formulated by public-health campaigners—universal provision of potable water, at standard and sufficient pressures, twenty-four hours per day—would not be met without a significant increase in investment, and therefore prices. However, the ability-to-pay of most prospective customers was below what the private companies needed to charge; network expansion was either insufficiently profitable or too risky (likely due to a combination of high capital costs and fear of public expropriation of infrastructure once completed).

Following this period of experimentation with both public and private provision, water resources and supply gradually became the responsibility of governments across the industrializing world.[21] This transfer was actively supported by lenders and donors in what came to be known, in the post–Second World War period, as the "development community." These organizations largely endorsed the municipal hydraulic paradigm of water management. As discussed previously (particularly in chapter 2), the rationale was a desire to provide sufficient quantities of water, where and when needed, such that economic growth could proceed unconstrained. The paradigm emphasized the importance of mobilizing water through large-scale hydraulic works and integrated networks, as a strategic resource to support agricultural modernization, urbanization, and industrialization.[22] A key justification of the state hydraulic paradigm was the presumed existence of multiple market failures (notably, natural monopoly and environmental externalities) together with the high capital costs, long infrastructure lifetimes, and relatively low rates of return associated with universal water supply networks.[23] This stance was not unusual, as the state played a key role as a facilitator of growth and promoter of technological progress across utility and resource sectors in many industrializing countries.[24]

Equally compelling was the fact that private water supply was not necessarily a highly profitable enterprise. Capital costs were extremely high compared to potential revenues, even when compared to other utilities such as gas. In some cases, companies were barely profitable, owing to a range of factors, including competition among privately owned water companies in the larger industrial towns and increasing average costs due to pollution or exhaustion of local water resources. In other cases, companies were profitable, but because of the risks stated earlier were unwilling or incapable of extending water supply networks to all customers over the entire urban area.[25]

The arguments regarding the advantages of government ownership and management of water supply systems were also important. The ability of municipal governments to achieve economies of scale through integrated monopolistic provision, to cross-subsidize water bills to poorer customers, to cross-subsidize network expansion (particularly of sewerage) through municipal taxes, and to raise cheap debt (often through programs sponsored by higher levels of government) was thought to be a significant advantage over the private sector. Moreover, the provision of water—as a nonsubstitutable element and essential for life and health—gradually came to be understood as a civic duty and a source of civic pride, supported by campaigns bringing together an alliance of medical professionals, engineers, and social reformers (such as the UK "gas and water socialists" mentioned earlier). Subsidized water became established as a basic entitlement of all citizens; private water vendors, and the practice of the poor begging or stealing water, disappeared from city streets in industrialized countries.[26] In nonindustrialized countries, water supply networks remained more limited in extent; any large-scale network expansion that did take place was almost always funded and managed by the public sector.

For all of these reasons, private sector activity in water supply began to decline by the late nineteenth century (table 3.1). In Britain, for example,

TABLE 3.1. *Public and Private Water Suppliers, United States (1800–1924)*

Year	Total waterworks	Public	Percent	Private	Percent
1800	17	1	5.9	16	94.1
1810	27	5	18.5	22	81.5
1820	31	5	16.1	26	83.9
1830	45	9	20.0	36	80.0
1840	65	23	35.4	42	64.6
1850	84	33	39.3	51	60.7
1860	137	57	41.6	80	58.4
1870	244	116	47.5	128	52.5
1880	599	293	48.9	306	51.1
1890	1879	806	42.9	1073	57.1
1896	3197	1690	52.9	1490	46.6
1924	9850	6900	70.1	2950	29.9

Source: Melosi 2000a, 36, 74, 120.

local governments controlled over 80 percent of water supply systems by the outbreak of the First World War.[27] A similar pattern was evident in the United States. At the same time, the increasing availability of public financing mechanisms (in the United States, the municipal bond market) permitted a rapid expansion in the number of public water supply systems.[28] With few exceptions, municipal ownership and management remained dominant for much of the twentieth century, in both developed and developing countries.

In summary, the nineteenth-century experiment with private water provision failed in part because of the inability of the private sector to profitably achieve the universal provision of water supply (with relatively uniform standards for water pressure and quality at low prices) that had become necessary in cities. Given the high costs of infrastructure, the activity of private companies outside of urban areas was limited by low rural population densities and affordability issues. Key risks for nineteenth-century companies were the low ability-to-pay of poor households, low profitability (particularly given the capital-intensive nature of network expansion), the costs of externalities (particularly those related to pollution, public health, and firefighting), and competition. The private sector disappeared or played only a minor role in most countries, and most water supply systems around the world were owned and managed by governments—colonial and postcolonial, communist and capitalist—for much of the twentieth century.

Two important exceptions were Britain and France. In Britain, small, private companies were permitted to continue owning and operating water supply infrastructure, but they were tightly regulated (with dividends capped, and non-water-sector investments prohibited). Private ownership of water supply infrastructure was expanded with the privatization in 1989 of all ten of the large, publicly owned water utilities in England and Wales.[29] In France, local governments continued to own the infrastructure, but private companies concentrated on the long-term management of water supply systems, supplying three-quarters of the French population. These two countries were to provide important models for the twentieth-century experiment with private sector provision in developing countries.

New Drivers for Water Privatization in the Late Twentieth Century

By the final decades of the century (see chapter 2), the conventional paradigm for managing water (the "municipal hydraulic paradigm") had begun to attract significant criticism. Large amounts of public capital had been invested, usually on a subsidized basis, in order to develop municipal water supply systems and large-scale hydraulic works, based on a desire to provide sufficient quantities of water, where and when needed, such that economic growth could proceed unconstrained. Three shortcomings stood out: (1) the failure to adequately address environmental externalities; (2) the failure to extend water supply universally (particularly in developing countries); and (3) the failure to invest adequately, as indicated by the infrastructure "deficit" (in high-income countries) and infrastructure "gap" (in lower- and middle-income countries). These "state failures," as they came to be known, were used in support of counterarguments to the long-held view that water supply systems were subject to multiple "market failures" (especially natural monopoly and externalities). In developing countries, the government agencies that managed urban water supply networks were often (although not always) beset by a range of problems.[30] This familiar litany (albeit contested by defendants of public water provision) included low coverage rates, low rates of cost recovery, low tariffs, underinvestment, deteriorating infrastructure, overstaffing, inefficient management, and unresponsiveness to the needs of the poor.

Over the past two decades, governments have attempted to address these issues through wide-ranging reforms of water supply management. As discussed in some detail in chapter 1, these reforms typically entail a triple strategy: (1) commercialization, through the introduction of commercial principles such as full cost recovery; (2) privatization, through a range of private sector participation contracts;[31] and (3) liberalization of governance, through broad-ranging water sector reform. Thus the privatization of water supply utilities is often accompanied by the introduction of new legal frameworks favoring cost recovery, framing water as an economic good, and permitting the creation of water markets. Other common reforms are the establishment of competition policies (sectoral reforms to allow private entry), the creation of regulatory frameworks and agencies, the requirement to tender a particular service or management of a specific utility to the private sector, and reduction in public sector jurisdictional fragmentation and overlap (table 3.2). These broader institutional and governance reforms set the stage for privatization; without them, privatization contracts are less likely to occur or to be successful.[32] Privatization, in other words, is part of a much broader transformation of water governance, explored in chapter 1 as a phenomenon of "market environmentalism," which many resource sectors have undergone over the past two decades. Privatization usually occurs along with other reforms that alter our sociopolitical, economic, and ecological relationships with water.

These reforms corresponded with a new set of policy priorities promoted by some development actors (particularly lenders and donors) in the 1990s. One aspect of these reforms was the set of economic policy preferences referred to as the Washington Consensus (discussed in chapter 2). This so-called (and short-lived) consensus was characterized by financial liberalization, privatization, deregulation, the creation of secure property rights, tax reform, the introduction of competition, and public sector fiscal "discipline."[33] This was closely articulated with the increasing integration and deregulation of global financial markets, enabling greater flows of private finance. The Washington Consensus emerged, in part, from fiscal crises experienced by governments of many lower- and middle-income countries in the 1970s and 1980s. Such crises were marked by high indebtedness, high inflation rates, and slow or negative growth. Critics argued that the debt crisis was an example of how misguided policies of multilateral financial institutions created "mal-development," and that the Washington Consensus aggravated rather than resolved the underlying problems facing developing countries.[34]

These reforms were also connected to the ascendance of the doctrine of sustainable development in environmental policy. Fierce critiques of the environmental blindness of mainstream bilateral aid agencies and multilateral financial institutions (notably the World Bank) in the 1980s had led to the "greening" of development organizations.[35] But the translation of this new environmental awareness into policy was shaped by the need to sustain

TABLE 3.2. *Water Sector Reforms (selected examples)*

Country	Year	Legislation	Provisions/policies
Armenia	2002	Water Code of the Republic of Armenia	National framework legislation with full-cost pricing provisions.
Azerbaijan	2001	Law of the Azerbaijan Republic on Water Economy of Municipalities	Transfers authority for "water structures" to municipalities; provides for cost-based water charges to "ensure efficient water allocation" and to finance system upgrades and development of new sources.
Burkina Faso	2001	Law 002/2001	Framework legislation for water resources management and water services with "polluter-pays" provisions and water charges to finance water sector.
Chad	1999	Law 15/PR/99: Water Code	Establishes cost-based tariff system; allows state to delegate public water service responsibilities to private sector.
Czech Republic	2001	The Water Act	National framework legislation including water use and wastewater discharge fees.
Estonia	1999	Public Water Supply and Sewerage Act	Authorizes connection charges to public water supply systems adequate to "develop the public water supply and sewerage system," "comply with environmental protection requirements," and "operate with justified profitability."
Finland	2001	Act 119 on Water Services	Requires that charges for water services "cover the investments and costs of the water supply plant in the long term," "promote the sparing use of water," and promote pollution reduction.
Indonesia	2006	Regulation 23/2006 (Law 7/2004)	Sets out full cost recovery and efficiency requirements. The new Water Law reinforces these provisions, mandates decentralization of control over Indonesia's water resources to regional entities and water user associations, and contains provisions allowing for private sector involvement.
Madagascar	1999	Law 98-029: Water Code	Establishes full cost recovery as a basis for water charges.

(continued)

TABLE 3.2. (*Continued*)

Country	Year	Legislation	Provisions/policies
Portugal	2005	Water Act 58/2005	Implementation of EU Directive 2000/60/EC of the European Parliament and of the Council establishing a framework for action in the field of water policy. Establishes principle of water as a scarce economic good, cost-based pricing, use of price mechanism to encourage highest-value use, and internalization of all costs.
Scotland (United Kingdom)	2003	Water Environment and Water Services (Scotland) Act (2003) and Water Industry (Scotland) Act (2002)	Response to an EU directive that "requires member states to take account of the need to recover the costs of water services as a way of encouraging the sustainable use of water resources." Replaces existing water and sewerage authorities with new entities having "a greater measure of commercial freedom."
South Africa	1994	National Water Act	Water pricing framework includes full cost recovery, user-pays and polluter-pays principles as a means of encouraging water conservation. Water use charges will be used as a means of encouraging reduction in waste. Provision is made for incentives for effective and efficient water use, and for penalties for nonpayment of water charges.
Spain	2001	Law 46/18999: Water Law	Establishes cost-recovery basis for tariffs on water from publicly financed projects.
Tajikistan	2000	Water Code	Establishes principle of fee-based water use; provides for water use fees to finance drinking-water supply system; explicitly prohibits privatization of drinking-water supply systems.
Tanzania	1994, 1999	National Water Policy and Water Law	Free water phased out and full cost recovery provisions implemented. Corporatization of country's largest water supplier, DAWASA, and provisions for private sector participation.

Source: Conca 2006; FAO FAOLEX database.

economic growth. Economic concerns and commercial commitments were thus integrated into environmental policy over the 1980s and 1990s, as reflected in the "liberalization" of international approaches to the doctrine of "sustainable development."[36] This produced, in turn, a new set of commercial approaches to environmental issues, largely supported by methods drawn from environmental economics. Water privatization was in line with this broader process of "market environmentalism" (first introduced in chapter 1). The Dublin Statement and Principles on Water, produced in 1992 in the context of the first global summit on the environment, capture the essence of this approach by including—along with traditional development concerns such as governance and gender—the controversial assertion that "water has an economic value in all its competing uses and should be recognized as an economic good."[37] Critiques of this approach were wideranging and included technical critiques of the impossibility of accurately valuing nature, ethical critiques regarding the treatment of nature as a commodity, and the voicing of spiritual concerns about worldviews incompatible with the commercialization of the environment.[38]

Privatization also found support among local and national politicians in some instances. The promise of instant cash from divestiture or signing bonuses was attractive, particularly for the many states that were facing increasingly serious foreign debt obligations. The potential for mobilizing "off–balance sheet" finance through private sector involvement was another selling point for indebted governments. The prospect of higher cost recovery, enabling repayment of (frequently nonperforming) loans back to international financial institutions, was in some cases an imperative. Other attractions included the possibility of devolving political accountability for unpopular reforms, such as tariff increases, and weakening the power of trade unions resisting those reforms. Another factor, in some cases, was the desire to escape requirements imposed by multilateral financial institutions regarding competitive bidding and procurement. Privatization of service delivery would also mean privatization of accountability, with less strict competition and tendering rules. Corporate confidentiality, for example, would enable contracts and key information to be removed from public scrutiny.

These drivers coincided with the initiative to internationalize operations on the part of water services corporations largely based in Europe (particularly France and England). With little room for domestic expansion, international opportunities beckoned in the 1990s. This was particularly the case for the recently privatized British water companies; lightly regulated, they had significant profits to invest (clawed back in the late 1990s by the British water supply regulator). The relatively high rates of potential profit (from

15 percent and upward in most contracts) were a significant "pull" factor. And the crisis in water supply access in cities of the South was suggestive of a potentially enormous market. The possibility of generating long-term, stable revenues (as with their domestic operations), while providing privileged access into lucrative public-works (infrastructure) contracts for their affiliates was another driver, particularly in the case of the highly diversified French water companies, with subsidiaries in businesses ranging from electricity to construction and engineering services. This coincided neatly with the availability of cheap credit (from private sources as well as from multilateral financial institutions) in the 1990s as a key enabling factor driving water companies to internationalize. In many instances, this was based on assumptions about the relatively low cost of borrowing money and the relative ease of access for these funds. The result was a rapid increase in the number of private sector participation contracts as well as the total financing directed toward private managers of water supply systems.

THE RESURGENCE AND PARTIAL RETREAT OF THE PRIVATE SECTOR

The early 1990s were promising for proponents of private sector participation in urban water supply. According to proponents, private sector participation in the urban water supply sector displayed an impressive increase, both in terms of numbers of projects and investment (figure 3.1). As with nearly every aspect of the privatization debate, the data on which figure 3.1 is based are disputed, and the World Bank has been accused of providing greatly overestimated statistics on private sector participation.[39] But our main interest here is in the discursive strategies that this data supported: predictions of a massive expansion of private sector activity in urban areas around the world.

This discursive story line proved, however, to be impossible to sustain as rates of growth in private sector contracts declined by the late 1990s. Private sector investment peaked in 1997. This trend was in part structural, as the Asian, Russian, and Argentine financial crises, as well as the bursting of the high-tech bubble, took their toll on foreign direct investment (FDI) flows.[40] However, whereas investment rebounded in some sectors, this was not the case for water supply: investment was significantly lower in the period 2000–2005 as compared with the previous decade, reflecting the trend toward fewer, smaller projects.[41] Cancellations across all utility sectors increased after 2000, but the highest rate of cancellations (on the basis of total investment) was in the water sector (figure 3.2).[42] Moreover, water attracted

3.1. *Water and sanitation projects with private sector participation, worldwide, 1990–2008 (data drawn from the World Bank PPI database)*

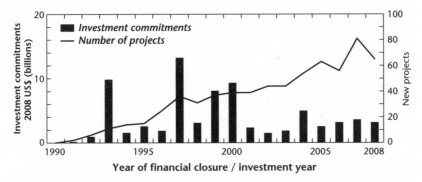

3.2. *Cumulative project cancellations, 1990–2005 US$ (data drawn from the World Bank PPI database)*

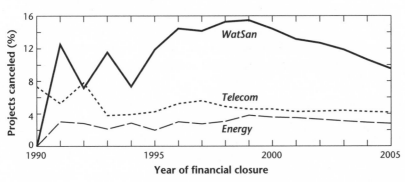

the smallest amount of investment of any utility sector.[43] This disparity is unsurprising, given the higher capital-investment requirements and higher risks in the water sector than some other utility sectors.[44] As the World Bank acknowledged in 2003, private sector finance had provided less than 10 percent of necessary capital, and the declining trend was likely to continue, exacerbated by the global financial crisis that had taken root by 2008.[45]

A Retreat from the "Concession" Model

Some observers argued that these trends implied a "retreat" of private companies. If framed in terms of population served, this view is incorrect: the total population served by private water operators continued to increase, albeit at a slower pace.[46] A more accurate interpretation would be that private companies began to show an increasing reluctance to engage in the long-term concession contracts that had become typical of the private sector participation projects of the 1990s.[47] Concession contracts were smaller and tended to be restricted to countries with access to long-term local-currency financing.[48] Private companies also reoriented their strategies toward shorter, lower-risk contracts that required little or no private investment, such as management contracts.[49]

Moreover, these observations are not universally true, as the level of private sector activity was spatially variable. Concession contracts were more rare in regions where overall investment was lowest (particularly sub-Saharan Africa), reflecting an investment bias toward east Asia (particularly China) and Latin America (figure 3.3).[50] It is worth noting that this spatial bias in investments is equally true of overall patterns of donor funding for water and sanitation, in which the lowest-income regions and countries tend to be disproportionately underfunded.[51] In other words, the pattern of private sector activity has, in recent years, mimicked the distribution of multilateral and bilateral aid funding for water supply.

Yet in middle-income countries, new "local" private sector firms emerged and began to hold an important market share in some areas (e.g., Brazil). Some argued that these trends implied a worrisome (but expected) strategy of "cherry-picking" on the part of private companies, whereas others observed that they were merely a healthy sign of the end of oligopoly, and of a refinement of corporate strategies.[52] The retreat of the private sector was, in this sense, partial: it is best understood as a calculated retreat from supplying lower-income consumers and regions. The result is that (contrary to hopes in the 1990s) private companies are not likely to mobilize significant amounts of finance for urban water supply networks in developing countries.[53]

These trends suggest that the mainstream model of water privatization (long-term, large concessions with international water companies) was not well suited to conditions in developing (and in particular low-income) countries. This would seem to endorse the views of privatization critics, who argue that universal provision was only achieved when the water sector was taken under public sector ownership and management, and that private sector participation models—largely developed in England and France—were unsuitable for achieving universal provision in developing countries precisely because of the large numbers of unconnected households with low ability-to-pay.[54]

Before we turn to the question of water supply for the poor, the reasons for the partial retreat of the private sector merit scrutiny. First, a series of issues (framed as "risks" by economists) threatened profitability. Relatively few cities were sufficiently large, with sufficiently developed water supply

3.3. *Water projects with private sector participation, by region, 1990–2008 (data drawn from the World Bank PPI database)*

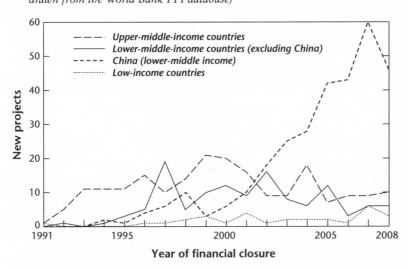

utilities and a sufficiently large middle class able to support cost recovery, to attract private sector interest. These had largely been cherry-picked early on. Many of the remaining contracts were characterized by low rates of return and longer payback periods than other infrastructure sectors. The need to deal with local governments ("subsovereigns") or multiple levels of government rather than solely national government (as for telecommunications and electricity) made contract negotiations difficult, particularly in cases of weak local-government capacity. Lack of effective state regulatory capacity was not originally a central focus on the part of donors. In fact, "regulation by contract" turned out to be time consuming, expensive, and highly conflictual, particularly with respect to tariff negotiations.[55] Renegotiations were particularly conflictual in the case of "opportunistic bidding," where the private firm winning the contract deliberately priced its original bid low, in the expectation of early renegotiation. Ironically, "state failure" was in this instance an argument against rather than for privatization. Indeed, this point can be generalized: weak government capacity can be presented as an argument against privatization (given the risks of weak regulation) rather than (as is frequently the case) an argument in favor. In retrospect, privatization contracts were more robust in the (relatively rare cases of) jurisdictions with strong regulatory frameworks (which had admittedly not been at the forefront of donors' and lenders' concerns in the 1990s).[56]

Many major water privatization contracts were also undermined by currency devaluations ("currency risk"), as demonstrated by the effect of currency devaluations in Latin America and Asia on high-profile privatization contracts in Buenos Aires and Jakarta. Such factors lent greater attractiveness to secure investments, with potentially (but realistically) lower rates of return, in Europe, North America, and middle-income countries with high growth potential (notably China). But growth in market share was in some instances limited, owing to several factors: public protests, organized through highly effective coalitions of unions, environmental groups, and citizen/consumer advocacy groups; the perception that privatization was politically problematic (particularly, in North America, following high-profile cancellations of private sector participation contracts in Atlanta, Hamilton [Ontario], and Stockton [California]); and the opinion that sufficient public funds and capacity existed in higher-income countries, where the greater cost of private capital outweighed the potential benefits of private sector involvement.[57]

Protests—both local and international—significantly raised political risks for private companies.[58] These protests included a variety of mechanisms: court cases, campaigns to pressure governments into canceling contracts, and street protests (in rare cases leading to contract cancellation or expropriation). This opposition brought together a broad range of groups, representing a range of interests, including organized labor, consumers, environmentalists, women's groups, and religious organizations.[59] In some instances, protests were framed in terms of an outright rejection of private sector participation; in other instances, specific issues were disputed, such as tariff increases. The views of protesters were, meanwhile, disputed by advocates of private sector participation. By 2005, even the World Bank acknowledged the fact that "the frequent (not inevitable) result [of water privatization] was popular protests, dissatisfied governments, and unhappy investors."[60] It seems reasonable to conclude that protests had the effect of raising the costs of private sector participation in some instances, and, in others, reduced private sector interest in long-term concessions (chapter 5 provides greater detail on protests over water privatization).

A third issue was "revenue risk," or the difficulty of recovering costs sufficient to achieve desired rates of profitability (which were often well over 10 to 15 percent). The political costs of raising tariffs to levels sufficient for cost recovery proved to be too great in a number of cases, given water's "necessity for life, positive health externalities, and history of being underpriced."[61] The perception of water as a "social good" was, as multilateral lenders and donors increasingly recognized, so powerful as to render politically impossible the necessary tariff increases, bill collection, and disconnection policies required by private sector involvement. This was particularly the case where the marginal cost of water increased steeply (implying large, rapid price increases) and where wastewater generated large externalities.[62] Given these constraints, private companies could not create the conditions in which profitability could be secured while simultaneously undertaking promised network expansions.

Finally, widespread and well-coordinated protests posed serious challenges to the public legitimacy of private companies. These ranged from an outright rejection of private sector participation in water supply to targeted critiques of privatization processes (the latter echoing those of the "mass" and "voucher" privatizations in the postcommunist transition economies, which by the late 1990s were the subject of outspoken critique, even among former proponents).[63] Public protest played a significant role in contract cancellations—or as activists sometimes call them, "remunicipalizations"—in several high-profile cases, perhaps the best known of which are La Paz and Cochabamba (Bolivia). One should not overemphasize the impacts of public protests, which were highly variable (with more occurring in Latin America, for example, than Asia or Africa). Nonetheless, the spectacle of protest attracted much international attention and may have had an effect on some renegotiation processes. Project cancellations on the ground, however, were not solely or even largely attributable to public protest; rather, these often stemmed from the underlying risks experienced by private companies and reflected the fact that the privatization model vigorously promoted by lenders and donors in the 1990s was not viable, for the reasons outlined.

Meanwhile, in wealthy countries, private companies also faced a series of challenges and highly publicized retreats. The reregulation of the British water supply industry following the election of the Labour Party in 1997 included a clawback of "windfall profits" on the order of several billion dollars. In France, high-profile private sector contracts were rescinded in two major cities, Grenoble (2000) and Paris (2008)—the former following the criminal conviction of senior company executives for corruption pertaining to the award of the contract.[64] Faced with low profitability in their international operations, British companies withdrew overseas operations, and only one remained extremely active on the international market by late 2008.

WATERING THE THIRSTY POOR?

The assertion that privatization could help the poor gain access to water supply (the "pro-poor" argument, in development jargon) may seem, at first glance, counterintuitive. Surely private companies demand profits and raise prices accordingly? This latter view would be refuted by many proponents of privatization. One of the prominent arguments in favor of private sector participation was the reasoning that extending networks, even on a full-cost-recovery basis, would enable the poor to pay less per unit volume of water, while improving service quality.[65] The logic is as follows: the extension of water networks to households without access would enable them to stop relying on expensive, poor quality alternatives (such as water vendors or private wells). Government water supply systems, in other words, discriminate against the poor through limiting networks to wealthier neighborhoods, which receive subsidized water. This is an obvious overgeneralization of a nuanced argument but is broadly representative of the "pro-poor" reasoning put forward in favor of privatization.[66]

New connections, of course, require new finance. When universal water supply networks were developed in Western countries, subsidies (from higher levels of government to municipal governments, and between consumers) were an important source of finance. But the logic of the market that dominates contemporary water policy requires cost-recovery pricing, which imposes the full costs of the system on the user. In the short to medium term, financing may "bridge" the gap in available funds, but ultimately users of the water network must pay for the service (although for some, the income support to maintain water bills at a certain level of affordability is also acceptable). In the absence of public sector financing, the private sector was, at least in the 1990s, a hoped-for source of much-needed financing for the water sector, particularly so that access could be extended to the poor.

This framing of the issues is disputed on two grounds. First, many who campaign for universal water access argue in favor of subsidies. In support, they assert that a minimum, universal level of water supply access (and sanitation) is necessary for human dignity (and, often, that it is a human right). Accordingly, subsidies from wealthier to poorer consumers (and, indeed, from wealthier to poorer countries) are required. Cost recovery is not a viable option given high levels of poverty in many urban areas. Often, these campaigners reject proposals that favor dual-standard systems (flush toilets for the wealthy, pit latrines for the poor), which are sometimes put forward by development organizations and community groups.

From this perspective, if universality is the goal, the issue of subsidies (via tax-financed public spending) provided by government networks to the wealthy is a red herring. Once everyone is connected, everyone benefits equally; because incomes are likely to be unequal, subsidies will have a significant, progressive effect.[67] The political choice to enact solidarity through state-led redistributive policies of this kind characterized the growth of welfare states in many Western countries; it may be fair to say that modern water supply systems are the outcome. To suggest that this strategy not be followed in developing countries (owing to lack of means or capacity) runs against the broader "moral economy" (to use Thompson's term) on which public services are predicated.[68]

Second, many dispute the assumption that private companies are likely to be of direct assistance in mobilizing finance.[69] The evidence, as we have indicated, suggests the contrary. The failure to mobilize finance was not, it should be emphasized, soley due to the concurrent financial crisis and global economic recession that had arisen by 2008. Rather, the multiple risks to which the private sector was subject undermined profitability and jeopardized access to finance. Moreover, the continued availability of concessional public finance, intended to foster private sector provision, reduced incentives to seek out private finance. Indeed, recent innovations in funding models and debates over water financing suggest investors are now more favorable to water sector–related investment opportunities that do not rely on the conventional PSP models of the 1990s nor solely channel funds through the water multinationals.

Thus, it is unsurprising that the record of new connections by private companies for poor customers is disappointing. The most recent large-scale survey, published in 2009 by the World Bank, estimates that private water operators have provided access (i.e., new connections) to piped water for more than 24 million people in developing countries since 1990.[70] But it is important to qualify this figure by noting that actual connections fell significantly short of original contract targets in most instances. Moreover, the bulk of financing for new connections came from public financing and tariff (operational) revenues—not new private sector finance.[71] Indeed, one of the few studies to have systematically examined the comparative performance of the public and private sectors in making new connections found no significant difference between public and private operators.[72] Because water is essential for life, both public and private utilities find it difficult to charge on a full-cost-recovery basis where there are large numbers of poor customers with low ability-to-pay. The failure of the "concession model" is actually just one example of a more generalized failure of conventional models of water services provision (whether under public or private management) to produce appropriate financing models.

PUBLIC VERSUS PRIVATE PERFORMANCE

In the early 2000s, the development community began to voice concerns about the "thirsty poor," and about the retreat of the private sector. New development goals had been formulated (such as the Millennium Development Goal of reducing by half the number of people without access to safe water by 2015).[73] At the same time, statements regarding the respective merits of the public and private sectors, adopting a more moderate tone than that of the 1990s, were issued by the World Bank and IMF, with the latter stating that "it cannot be taken for granted that PPPs are more efficient than public investment and government supply of services."[74] This carried over into policy documents. For example, the emphasis on consultation, building stakeholder consensus, robust regulation, and an objective assessment of the appropriate role and potential contribution of the private sector in the World Bank's second edition of its "toolkit" for private sector participation in water supply contrasts dramatically with the first edition, which gave little or no attention to these issues.[75] But support for privatization—and for the broader scope of commercial reforms of the water sector—remained largely intact. Many proponents continued to argue that, under the right conditions, private sector contracts could still outperform public management.

The question of comparative performance has been central to the privatization debate. Despite a decade and a half of experimentation, the question of the relative performance of the public and private sectors remains far from settled. Most of the debate has centered on the relative merits of the public and private sector in managing large-scale reticulated water-supply networks (excluding community cooperatives and smaller systems, which have been largely viewed as insignificant or anachronistic, despite the large numbers of people that they supply). Unbiased research is rare; an examination of comparative performance is often influenced by ideological commitments. It is thus somewhat predictable that proponents and opponents of privatization rarely agree on research strategies—on the relative value, for example, of cross-sectional econometric analysis versus detailed case studies. In addition, studies have been plagued by a lack of data, particularly with respect to reliable baseline information to enable the evolution of parameters. The choice of parameters is also important (efficiency, connections, tariffs, affordability, etc.). Efficiency tends to be the primary or even sole focus of studies by academic economists.[76] Opponents and proponents of PSP have tended to select different parameters, thereby producing competing (and often divergent) evaluations of the same contracts.

With these caveats in mind, there are two main positions adopted within the "comparative performance" debate. The first suggests that whether an operator is public or private does not have a strong effect on management outcomes. For example, studies of the performance of water supply utilities in Asia, the United States, the United Kingdom, and Europe suggest that ownership (public or private) does not predict the efficiency of water service providers.[77] More generally, the public administration literature states that there is no association between the formal-legal status of a public service provider and its service delivery performance.[78] Moreover, it is a fact that goals vary under public and private management—the latter prioritizing profitability and, if appropriately regulated, economic efficiency, the former prioritizing other goals such as security of supply. This reality confounds attempts to analyze changes in performance solely on the basis of efficiency.[79] And private companies, indeed, have goals other than efficiency—namely profit maximization. One recent World Bank paper claims, for example, that private water supply contracts generate gains in labor productivity (reductions in employment per connection or unit output) but not lower prices or higher investment, raising the possibility that "the private operator may reap all the gains through profits, passing on none of the cost savings to consumers."[80]

It is this sort of research that lies behind statements like this one made by the International Monetary Fund: "Although much of the case for PPPs rests on the relative efficiency of the private sector, [and despite the] extensive literature on this subject, the theory is ambiguous and the empirical evidence is mixed."[81] This point of view tends to emphasize the importance of institutions (in the sociological sense of rules, norms, and customs). Inadequate or absent regulation, for example, is often the focus of this literature, which argues that poor regulation contributes to poor performance—on the part of both public and private operators.[82]

The second position holds that public or private management does have an effect on performance, and that this is attributable to the ownership of the operator. Some argue, for example, that private sector involvement improves performance through higher efficiency and cost recovery, enabling additional sources of finance or higher connection rates for poor households.[83] Critics argue that studies demonstrating positive outcomes of privatization are subject to certain biases: for example, they fail to account for the self-selecting nature of the privatization process, in which only the best-performing firms (large, and potentially or actually profitable) are likely to be privatized. Others argue that private sector participation negatively affects performance through raising the cost of capital, reducing long-term investment in infrastructure repair and replacement, increasing corruption, or reducing affordability due to tariff increases.[84] These divergent interpretations may be due, of course, to ideological bias; opponents (or proponents) of privatization may be more likely to focus on negative (or positive) effects of involving the private sector, thereby emphasizing the importance of ownership. But they agree on one central point: the ownership of the operator (whether public or private) is significant for how a water utility actually performs.[85]

Querying Commercialization

It is probably fair to say that the actual effects of public versus private management on specific performance variables, such as efficiency, are unclear.[86] But it would be untrue to conclude, as some do, that public and private operators are no different from one another. The municipal hydraulic model (as discussed in chapters 1 and 2) differs significantly from the "market environmentalism" (or commercial) model on a number of points, implying two very different sets of institutional approaches, and hence differences in performance. It is on this basis that protests against water privatization have broadened into protests against commercialization in general.

One way out of the privatization impasse is a compromise: combining the best of the public and private sectors. A simple example of this is the approach used in the Netherlands, where water supply utilities are run like private corporations. The Dutch model is one in which private businesses operating on commercial principles have public owners. This is distinct from the French and British models of privatization and regulation that dominated the debate in the early 1990s.[87] The Dutch approach holds that water must be commercialized in order to be sustainably managed. Some of the key elements of this approach include full-cost-recovery pricing, the treatment of water as an economic good that must be allocated to its highest-value uses, and the definition of users as "customers," with the attendant claims on the business that provides them with water services. But the viability and impacts of commercializing water supply are contested.

Proponents of public water, for example, reject the assumption that water services must be self-financing. Given that water supply systems are extremely capital intensive, they argue that affordability concerns are almost inevitable for some segment of the population. In order to avoid "water poverty" (where the amount spent on water goes beyond a threshold of income), even in well-developed systems in high-income countries, some level of cross-subsidization or income support is necessary. In lower-income countries, where networks need to be extended, a second critique arises. Here, the critique pertains to the assumption that users must fully reimburse all of the costs of building (as well as maintaining) systems.[88] Proponents of public water supply management often make reference to the historical development of water networks in high-income countries, arguing that universal provision was achieved only when cross-subsidies (between consumers and between different orders of government) were implemented. The need for subsidies of some sort is almost inevitable if the goal of significant network extension—much less universal provision—is to be realized.

This, in the view of proponents of public water, is a strong argument for bringing the water sector under public sector ownership and management. "Public" finance, in this sense, does not just mean sourcing funds from the public sector (e.g., through government bonds). It also means creating another model of water governance, whereby cross-subsidies (and a certain degree of social solidarity) are accepted. In other words, public water, according to its proponents, suggests a vision of solidarity among users (and eventually between users and the environment) that is radically different from that of the commercial model. Until this occurs, proponents of public water argue, the thirsty poor will remain unserved.

For this reason, proponents of public water often reject the suggestion that small-scale, informal private providers—the archetypal water vendor supplying slum dwellers with hand-hauled containers—can "fill the gap" left by the retreat of large private water companies. Interest in these "small-scale-independent providers" began growing from 2000 onward, with a more nuanced appreciation for the broad range of private actors providing water in urban areas: water vendors; property and land developers; community-based organizations and NGOs; and small engineering firms providing on-site, smaller-scale technologies (e.g., Solo, 1999).[89] These actors should be enrolled, proponents of private provision claimed, in "public-private community partnerships."[90] At times, the term "private" is (perhaps unhelpfully) extended to include all nonstate actors, including the broad range of "community" approaches that have long existed in water supply management, such as water cooperatives. Critics argue, however, that poor service quality, the loss of economies of scale, and the entrenchment of dual or multiple water-supply delivery standards associated with socioeconomic status were serious concerns. The reasoning was that these "reformed" private sector approaches merely endorsed an unsatisfactory status quo, which misguidedly overlooked the necessary role of governments and public policy to correct market imperfections and provide water for all, on an equal basis.[91] Accordingly, some critics tend to reject reformed private sector approaches that include informal private actors.

Proponents of public water, in other words, continue to advocate the vision of the modern integrated ideal described in chapter 2: the provision by government-run networks of universal access to affordable water supply, as a material emblem of citizenship. They advocate a renewed focus on supporting public water utilities and point to comparative research supporting the existence of well-managed public utilities in the developing world.[92] Some of these approaches have received institutional support, such as the public-public partnerships promoted through the UN Secretary-General's Advisory Board on Water and Sanitation or the renewed consideration by lenders and donors of fostering capacity of public sector utilities and public financing mechanisms.[93]

Proponents of privatization argue, in contrast, that the flaws of government management are still persistent and widespread, and that water privatization is still a viable response. Some argue in support of management and lease contracts as a means of mobilizing private sector management expertise to improve performance, rejecting criticisms that the reduction of risk for the private sector also reduces performance incentives, while acknowledging that these contracts do not provide mechanisms for private financing of network extension.[94] Instead, they advocate new strategies of deepening private sector support through, for example, "political risk" coverage or new financing models.[95]

Is Water a Public or Private Good?

In this sense, the term "private" has undergone a subtle, but important, redefinition within the water privatization debate. It no longer solely implies the introduction of the formal private sector and a transfer of assets, whether temporary or permanent; rather, it has been expanded to include the informal private sector, whose assets have never come under state or corporate control. Moreover, it incorporates a broader range of the continuum of service provision (as evoked in figure 1.1) that exists in cities in the global South. It implies, in other words, that service provision is, in fact, a complex admixture of formal and informal, government and private sector activity.

In the absence of government services, a plethora of alternative strategies of service provision arises. Some of these involve illegally tapping into existing networks, as with the midnight plumbers and comrade electricians operating in South Africa's townships. More frequently, unregulated private alternatives, usually operating on a small scale and using artisanal technologies, spring up. Some of these services may be provided by NGOs, whereas others may be provided by individuals or small businesses for profit. More rarely, communities come together to provide themselves with the services that their governments are unable or unwilling to provide: cooperative water supply systems and sewerage networks, or self-built housing. All of these are "private"—in the sense of nongovernmental—approaches to services provision undertaken by members of the "public."

Is water an economic (or "private") good or a public good? The answer to this question is "neither." Water's status is ambiguous. This is most obvious in urban areas of the lower- and middle-income countries, where public and private sector provision of water supply typically overlap—and often compete. But it is also true in high-income countries, where most water supply delivery models are organized along a continuum from public to private. Few water supply systems are either purely public or private. Water (as with many other goods) thus presents us with a conundrum: it is itself neither public nor private, and our conventional models for service delivery (government and corporate) have proved unsatisfactory.

This position is at odds with much of the academic and policy literature on water supply, which has until recently largely overlooked the mixed public-private nature of water supply systems. Rather, emphasis has been placed on the relative merits of ideal-type public-versus-private delivery models. This focus has obscured important institutional dimensions of water supply provision, such as governance—understood not in a narrow technical sense but as a broader issue of politics and social power.

The public-private dichotomy is misleading for another and perhaps more fundamental reason, which undermines some of the core assumptions of the policy debate. This critique rests on the definitions of the terms. Our conflation of "public" with "government" and "private" with "market" obscures a wide range of social activities that fall beyond the sphere of states and private corporations. Arendt's "public realm," Habermas's "public sphere," and, more recently, Chatterjee's "political society," and Putnam's "social capital" are some of the more influential attempts to formulate the concept of what, in policy terms, is commonly known as "civil society."[96] Here, public and private are inextricably mixed, as in Habermas's formulation of the *public* sphere as a grouping of *private* people uniting (beyond the aegis of the administrative state) around matters of common concern.[97] From this perspective the term "public" is defined in terms of political community and citizenship and is analytically distinct from *both* states and markets.[98]

These examples challenge our definitions of public and private. The great bulk of the public policy debate is structured on the assumptions of utilitarian liberalism, in which the distinction between public and private equates to that between governmental and nongovernmental. Conventionally, the latter category is held to be synonymous with market (that is, for-profit) activity, although recent debates, particularly over the existence and importance of "social capital," have extended this term to include not-for-profit groups such as NGOs. From this perspective, the distinction between "public" or state authority and "private" individual activity is assumed to be clear. Debate over the scope of state and market activity is thus essentially a debate over jurisdiction, adjudicated via utilitarian standards of performance.

Reducing the debate to one of technical terms or managerial strategies is, however, misleading. Formal privatization—the focus of this chapter—is a process of expansion of the market, which creates new property relations and in so doing generates new society-nature relations.[99] But this occurs in the context of (and is often disrupted by) "alternative" technologies, governance strategies, and community economies of water that exist in many places around the world.

Conclusions: The Debate over Private Sector Participation at a Stalemate

By the end of the first decade of the new century, the international debate over water privatization was at a stalemate. Despite being vigorously promoted and supported by multilateral lending organizations and some bilateral aid agencies, private sector participation in water supply in developing countries failed to achieve some of the key results promised by proponents. An increasingly broad-based consensus had begun to develop by the middle of the decade that the role of private capital in urban infrastructure would be more limited than had been expected. But whereas some continued to argue for private sector–led models, others argued that a renewed public sector effort was necessary to extend the provision of urban water supply services to poor households.[100] In short, although there was a degree of consensus on the failure of the concession model, and on the partial retreat of the private sector, there was little agreement about the means through which "water for all" should be achieved.[101]

Discontent with the results of private sector participation in water supply among former proponents gradually began to stimulate debate on both the privatization experience and alternatives. These questions were typical of the debate over the so-called post–Washington Consensus (or postneoliberalism)[102] characterized by shifting definitions of public and private and the proliferation of alternative management models.[103] Could reformed and improved private sector models work? Should attention be focused on improving the performance of public utilities? Or were other models—financial, governance, and technological—required?

The debate over alternatives is the focus of part 2. One of its central concerns is the reformulated notion of public and private explored in this chapter. But to ease into the new approaches, I present in chapter 4 a case study of one city—Jakarta—which provides a compelling example of the issues.

4

CITIZENS WITHOUT A CITY

THE TECHNO-POLITICS OF URBAN
WATER GOVERNANCE

With Michelle Kooy

Shortly after the Asian financial crisis and the fall of President Suharto that rocked Indonesia's economy in 1998, a new commodity appeared in Jakarta's slums: bottled water. For years, middle-class residents of the city had relied on bottled water for drinking, because the government-run network covered only a third of the city, and the water it supplied was unfit to drink. Privatization of the water supply network that same year had failed to result in hoped-for new connections and better water quality.

But as massive job losses and inflation hit, prices for basic goods rapidly spiraled out of control, and "name-brand" water became too expensive for most households. Instead, Jakartans began flocking to the new, private water "microtreatment" plants that had sprung up around the city, selling purified water at less than one-quarter the price of brand names. As competition increased, prices dropped, and the number of private microtreatment plants—completely unregulated—quickly grew. Soon the urban poor—long used to fetching and storing their water in buckets and gallon drums—also became avid customers. Slum residents began scavenging garbage dumps for empty water bottles that they could refill at air isi ulang *(water refill) depots.*[1] *Wealthy households continued to drink brand-name water flown in directly from France, but this had lost some of its cachet. Bottled water had been democratized.*

What are we to make of Jakarta's thriving bottled-water economy? Neoliberals might celebrate the creativity of the market, rapidly growing to meet new demands, entirely outside the regulatory reach of the state. Environmentalists might be caught between condemning water pollution and (somewhat perversely) praising the thrifty diversion of plastic bottles from the city's dumps. Development and aid workers might focus on the health benefits and risks for the urban poor, while noting that relatively high prices and a lack of regulation raise questions about long-term sustainability. Antiprivatization and alter-globalization activists would undoubtedly condemn the infiltration of the capitalist economy into yet another facet of urban life, pointing out the social, economic, and ecological costs of the reliance on bottled water and insisting on the need for an effective publicly owned and community-controlled water supply network.

In considering these different responses, we must remember that "public" services in many countries are limited to the elite. As described in this (and earlier) chapters, governments often lack the resources to provide all citizens the material emblems of citizenship (like electricity or water) on which their legitimacy is predicated. This is not just a question of financial resources but also of political entitlements. Social ruptures within postcolonial states have frequently meant that only a subset of the population are identified as full citizens, with the full set of rights and entitlements that political theory suggests they should hold. Water supply networks (like other infrastructure networks) reflect this fragmentation: they are archipelagos of elite services in the wider urban fabric.

How and why do water supply systems come to be fragmented? In this chapter, we attempt to answer this question through a case study of Jakarta, the capital of Indonesia. In essence, our argument is that both governments and private providers fail in important (sometimes similar, sometimes different) ways. This position is at odds with the conventional positions taken in the water privatization debate, in which blame is usually attributed to either public or private providers. Critics of government cite, for example, inattention to urban water supply issues by political elites, or lack of public finance for investment, or inefficient public-sector management of the water supply utility. Criticisms of private providers argue that the drive for profit fosters fragmentation as a means of creating lucrative niche markets; this implies that universal, equitable provision is unlikely when private providers are involved. This criticism of private sector provision is usually associated with a defense of public water supply systems. Usually, one explanation is understood to exclude the other.

In contrast, we suggest that both these arguments are, to an extent, valid; both public and private providers contribute to the fragmentation and differentiation of water supply access within cities. To understand how this might be the case, this chapter presents a case study of one city, over time. Of necessity, the chapter presents a historical perspective. Water infrastructure networks have one of the longest turnover times of any dedicated utility infrastructure; in many cities around the world, it is common to find pipes over one hundred years old still in service. Given this longevity, water networks physically embody successive phases of management and investment; historical choices thus shape contemporary constraints on water supply network managers. It follows that understanding contemporary patterns of fragmentation requires an analysis of the social and technical choices made over time regarding the expansion and distribution of a city's water supply system.

As illustrated by Jakarta, inequitable access is often literally hardwired into water supply networks. Scarcity, in other words, is mediated by human action; it is socially constructed but nonetheless very real.[2] As documented below, the production of scarcity (via exclusion from water network access) by both public and private actors is integral to the process of modernization and urbanization. The production of fragmented networks (or "archipelagos") is, in other words, no accident.

This is not to say, however, that fragmentation and exclusion are deliberate in a simplistic sense (although this is indeed sometimes the case). Rather, fragmented networks are perhaps better viewed as "unintended consequences of development" (to use James Ferguson's language).[3] That is, they are unintentional but nonetheless coherent outcomes of technical practices, political commitments, and cultural norms.

Our inability to recognize this coherence stems in part from a depoliticization of development. In her study of development schemes in Indonesia, Tania Li has usefully termed this process "rendering technical."[4] Similarly, Timothy Mitchell uses the term "techno-politics" in his study of the interrelationship between colonial power, technique, and expertise (the "rule of experts") in Egypt.[5] Both are making a similar argument: the depoliticization of development renders the process of exclusion normal, even banal—and difficult to *see* as a politicized process.

By this, we do not mean to imply any slavish imitation of Mitchell's approach. Rather, our characterization of this chapter as techno-politics refers to the same conceptual commitments invoked by Mitchell: his assertion of the interrelationship between a specific worldview (modern, secular rationality) and a specific set of techniques (both technological and social); his insistence that we must situate this worldview and its associated techniques in historical and cultural context; his claim that economic categories (and indeed the economy itself) are cultural constructions; and his consequent, powerful politicization of apparently technical, benign interventions, whether by colonial powers or development agencies.

The term "techno-politics" also serves to remind the reader that politics are materially expressed, in this case, via technologies of water supply infrastructure. The distribution of infrastructure networks in contemporary cities like Jakarta can be interpreted as a political ecology of exclusion via the production of urban nature (not the enclaves of green that tend to be the focus of urban studies, but of the physical, bodily "stuff" that circulates through cities: excreta, water, wastes).

Infrastructure also has, of course, what we might term a social dimension. In part, this is a question of the political economy of property relations. Land rights and tenure norms underlie the deeply unequal allocation of water and sanitation in modern cities, implying a link between property rights and other sorts of rights, including citizenship rights (and human rights as developed in chapter 5). Without secure tenure, for example, households often have difficulty accessing public services. But the question of tenure is not just a legal issue; it is frequently tied up with questions of ethnicity, identity, and class whereby different types of urban residents are constructed as distinct inhabitants of the city. Thus, our use of the term "social dimensions of infrastructure" also refers to the cultivation of tastes, norms, and water-use practices intimately connected to the construction of subjectivity, "ethnic" identity, and cultural commitments.

This implies a dual analytical task: the excavation of the political ecology of urban water, and the repoliticization of the story of Jakarta's water supply. In what follows, we resuscitate the city's history in order to explore how bodies, subjectivities, politics, urban space, and water infrastructure are mutually constitutive; to demonstrate the deeply political dimensions of lack of water access; and to uncover biases in the ostensibly technical, neutral process of building water supply networks. It is to this story—of the beginning of Jakarta's water network—that we now turn.

WATERING JAKARTA

The disparity in access to water supply in Jakarta is stark. Official estimates of the proportion of the city's population with water supply network connections in the home ("household connections") range from 46 to 56 percent.[6] Data from Indonesia's census suggest that only somewhere between 10 percent and 15 percent of the urban "very poor" (with monthly incomes of less than US$80)[7] have access to piped water supply in the home. But these figures are imprecise at best. Unofficial estimates, which attempt to account for the large number of informal residents in the city, suggest that less than a quarter of the true population has access to water from the network.[8] Domestic water consumption is estimated to be between seventy and eighty liters per person per day—one of the lowest in Asia.[9]

Those not connected to Jakarta's municipal water supply system rely on a variety of sources: shallow and deep groundwater; surface water via the municipal water-supply network; spring water; and (less frequently) direct use of rainwater, rivers, and streams. They also make use of a variety of distribution methods: household wells or rainwater collection systems; water vendors; bottled water; public standpipes; small, private networks connected to deep wells; and water trucks.[10] As with other large cities in developing countries, many of these "alternative" water-supply methods are more expensive, per unit volume, than piped water supply. As a result, many poor households pay a much higher proportion of their income for water than wealthier households do. Moreover, the quality of most of the water sources used by low-income households is very poor. Wealthier households with a networked connection, in other words, not only receive cheaper water but also spend much lower proportions of their income for much greater quantities of water of higher quality. This pattern is typical of many cities in developing countries.[11]

Jakarta's small municipal sewerage system connects less than 2 percent of households, a legacy of government policy treating sewage as a "private concern."[12] The majority of wastewater is disposed directly into rivers, canals, or to (often poorly functioning) septic tanks. The resulting contamination of surface water sources by sewage and industrial effluent is exacerbated by the lack of an effective urban waste collection system, with household waste collecting in canals that provide water sources and flood drainage for the city.[13] Open canals, largely conduits for sewage, regularly overflow into city streets during the rainy season.

As a consequence, poor water quality in the piped network (partially reliant on surface water sources within the city), and in shallow groundwater (the source for the majority of the city's poor residents) is of particular concern. Contamination by wastewater and industrial effluent, as well as salinization—purportedly due to seawater infiltration resulting from overpumping—have polluted Jakarta's shallow aquifer, the sole source of household supply for many poorer families in many areas of the city.[14] Rivers and canals are generally too polluted to use even for washing clothing.[15] Nor is the water delivered through the network potable, owing to poor quality of the raw water available to the water supply utility and poor infrastructural quality of the distribution network, leading to infiltration and contamination in the distribution mains after water has left the treatment plant. Indeed, public-health studies repeatedly find fecal coliform contamination in Jakarta's networked water-supply system, and residents are advised to boil their water.

The public-health impacts of this situation are predictable and have been well documented: high rates of water-related diseases, including gastrointestinal illness due to contaminated water, and parasite-related illnesses due to poor drainage, particularly in poorer areas.[16] Results of the first community-based surveillance study of diarrhea in Jakarta indicate that diarrheal diseases, many of them water related, are the third leading cause of morbidity in the general population and the leading cause of morbidity in infants.[17]

PRODUCING NATIVE SUBJECTS: THE COLONIAL WATER SUPPLY SYSTEM

Decades of development interventions by international donors and the Indonesian government have sought to reduce or eliminate the disparity of access to water supply in Jakarta, but differences in access have only deepened. The explanation for this apparent paradox begins with the colonial origins of Jakarta's water supply system.[18] Intended only for the small minority of European residents living within a small central area of the city (then known as Batavia), the water supply network drew on artesian springs to provide free water to the Europeans, while deliberately excluding "native" residents. In contrast to the surface water from the city's *kalis* (canals), on which the majority of Batavia's residents relied, European households now had access to scientifically monitored groundwater, circulated through iron pipes rather than via the pedestrian water vendors who moved constantly through the neighborhoods.

When connections were finally permitted for "native households," they were based on a dual design standard: 140 liters per day were to be distributed to European households, in contrast to an anticipated delivery of just sixty-five liters to native households, only a third of which would have direct access within the home.[19] The obvious result was a differentiation of access and consumption levels. In the 1930s, forty years after the water supply system had been constructed, Europeans (only 7 percent of the population) consumed 78 percent of the volume of water supplied to residential customers.[20] Indeed, "natives" were not provided with access to household connections until the late 1920s; most native households relied on mobile water vendors, who charged twice the cost per unit volume paid by tap water customers.

Why would the colonial government have adopted this strategy, particularly given Batavia's reputation as the "graveyard of the East" (which evoked the epidemics of cholera and typhoid that regularly plagued the city)? As elsewhere, new scientific theories of disease transmission via bacteria had begun supplanting formerly widespread notions of "miasmic" contamination. Disseminated largely by military doctors in Batavia, these theories proved congruent with the Dutch East Indies colonial government's evolving ambitions for its strategic colony, central to which was a new valorization of "modern European life" as the colonial cultural developmental ideal.[21] Given this confluence of factors, it seems that the colonial government would have embraced the concept of clean water for all city residents.

This ideal, however, was undermined by a growing *Indische* population created by immigration and intermarriage. This culturally distinctive population—the product of a century of intermarriage between Dutch settlers and indigenous families—lacked clear biological markers of distinct racial identities.[22] The policing of the blurred distinctions between "European" and "native" residents was a constant point of concern for the colonial administration from the mid-eighteenth century onward.[23] Long-term residents and "degenerate" Europeans[24] were problematized in colonial reports for their overfamiliarity with native culture and practice of unacceptable *Indische* traditions[25]—including their continued preference for traditional, untreated water sources and "unhygienic" water practices. *Indische* households relied on wells in the inner courtyard of the house, bathed using a *mandibak* (a traditional, manually filled shower basin) or in one of the many *kalis* (canals) running through the city, and drank unfiltered water. In contrast, "newer European homes," built by the wave of middle-class, newly arrived, solidly "Dutch" residents, were equipped with indoor bathrooms that enabled water-intensive hygienic practices to be carried out daily, in private, with filtered tap water also used for drinking.[26] Contemporary commentators and government campaigns repeatedly emphasized these "European" practices as modern, hygienic, and morally proper—in contrast to the widespread use of canals and wells. In practice, categories were much more confused; the existence of Europeans who chose to adopt native ways and live in the kampong (literally, "village"), for example, was a source of ongoing concern for the colonial administration. The definition of "European" practices should thus also be understood as a practice of policing the boundaries of identity, in a context where intermarriage and cultural interchange threatened to break down the cultural cleavages so central to colonial authority.

Within this context, an explanation of the Dutch government's decision not to supply water to the whole city becomes more intelligible: it may be read as a coded signifier identifying those of a racially pure European identity. The colonial official who conducted the first extensive (and scientific) study on the various qualities of water supply in Batavia, for example, argued that traditional methods of water treatment—that is, *lekstonen* (a simple lime-sandstone filter) and *martevanen* (large water pots for "settling" of particulate matter in river water)—and quality surveillance (largely through individual assessments of color, odor, and taste) undermined the morality and civilized status of the city's European residents. Europeans were instructed to replace their traditional *lekstonen* and sensory assessments of water quality with scientific assessments of hygienic requirements to address the unseen microscopic threats whose recognition was a marker of civilization.[27] Bathing in canals and drawing drinking water from rivers were similarly denoted as "degenerate" and unhygienic. Blame for cholera and typhoid epidemics was increasingly directed at natives' "hygienic circumstances."[28]

In this context, it is important to emphasize the material dimensions of water supply infrastructure: notably, the increased spatial segregation of the city enabled by the water supply system. Following its creation, European residents moved from the decentralized riverside garden villas of *Indische* society to a more concentrated cluster of European residences, deepening the contrast between the "civilized" urbanites of the city core and the still largely rural lifestyles of the city's kampung neighborhoods.[29] In the mid-1920s, one visitor to the city observed that "Batavia is a European town built by Europeans, *except* for the natives freely bathing in the canals and rivers."[30]

Indeed, the *kalis* were a source of particular concern for colonial chroniclers. Batavia's canals were "scenes of considerable activity from early morning till late afternoon. The banks of the canals, the steps that lead down into them, and the bamboo rafts are all crowded with throngs of half-naked native washerwomen and laundry men," who preferred "bathing, washing and defecating in streaming [rather than still] water."[31] Colonial hygiene officers and engineers considered this a public display of "their undeveloped approach to the functions of life" and of their ignorance of the "proper forms of urban life."[32] Observing native residents using the city's canals and rivers for bathing and laundry, contemporary writers described canal-bathers as "children" in both their practices and mind-set—natives seemingly had an utter "indifference to the sorrows and realities of life."[33] Conveniently for the colonial government, of course, this demonstrated how the "children" would continue to require the guidance of their colonial parent.

But as the development of the city's urban water supply infrastructure continued to parallel the patterns of European urban development in

Batavia,[34] the growing contrast in development and lifestyle between the European and native areas of the city began to embarrass the colonial government. In the 1940s, the government's Visman Commission recorded the extremity of racial segregation in a city composed of "two different worlds."[35] The water supply network was mobilized as a symbol for anticolonial criticism, as campaigners argued for a more equitable distribution of clean water to reduce native mortality rates and disease.[36]

Yet projects to improve native hygiene were surprisingly (at least to the colonial authorities) met with resistance. Network extension, through the building of communal standpipes, was met with vandalism and the widespread refusal of kampung residents to pay for water.[37] This, in turn, stalled further extension of the network, intended by the colonial authorities to be run on a cost-recovery basis (unlike the free connections available to Europeans). And only a minority of residents used the communal (yet enclosed) washing and bathing facilities constructed to "keep the [native] population out of the canals."[38] These facilities established private spaces for domestic functions and simultaneously differentiated the waters in which it was "proper" to wash oneself, one's clothes, and one's household items.[39]

Perversely, however, "native" households often had no choice but to use canal water. The colonial government continued to limit the extension of piped water supply into native areas of the city, despite repeated calls to extend the network, given the visible disparities between living conditions for European and native residents of the city. This was also because native residents were considered by some colonial authorities to have less need for access to the city's piped water supply infrastructure. A former officer of the Royal Engineers in the Dutch East Indies appointed by Batavia's Municipal Council in 1922 to review the designs for the city's subsequent water supply system[40] calculated the water needs of the city based on the different water-use habits of distinct racial groups: Europeans would use 150 liters per person daily, while Chinese/Arabs would use one hundred liters, and the "native" population would use only fifty liters.[41] Natives were not intended to be modern, nor European—and this distinction was simultaneously embedded into the city's water supply system and urban form.

STATE FAILURE: THE POSTCOLONIAL WATER SUPPLY SYSTEM

The image of the "native in the canals" continued to serve as a justification for government action following independence, which ended the armed nationalist struggle to oust the Dutch in 1949. President Sukarno's global ambitions as leader in the nonaligned movement (most notably through his leadership of the Bandung Conference in 1955) and national ambitions of developing and unifying a multiethnic, multilingual, and still tentative federation converged through Sukarno's personal project of modernizing and beautifying Jakarta.[42] However, the public monuments and highly visible infrastructure projects initiated under Sukarno's direct oversight were visibly undermined by the "masses of poor" residents and their "public striptease" of washing and bathing in the city's canals, compromising its identity as a "beacon for the third world."[43] Residents with "backward urban lifestyles" who did not contribute to making Jakarta a "Beautiful, Orderly, and Dynamic city" by adopting the "good norms and standards of city life"[44] were gradually—and sometimes forcibly—displaced to the nonserviced urban periphery.

In the decades following independence, water supply in Jakarta was characterized by many similarities to colonial patterns of water provision, although Dutch administrators and residents had long since fled the city. In part, this was because of technical constraints: surface-water supply continued to be distributed through the colonial piped network and was thereby restricted to more affluent areas of the city. But this is only a partial explanation, as major investments were made in the first decade following independence to both rehabilitate and expand the water supply network.[45] Networks to distribute this water were not, however, extended universally across the city, despite the fact that additional water supply was available; the network distributed less water than the treatment plants produced.[46]

This situation might seem curious, particularly given Sukarno's training as an architect and his ambition to convert Jakarta into a symbol both of Indonesian unity and of the development potential of newly independent postcolonial states. But investment in the city was directed primarily to showcase projects.[47] Water treatment and distribution, as "invisible" infrastructure, was a relatively low priority for the newly independent government dedicated to modernizing services for the elite, rather than universalizing services for the masses.[48] Network expansion was deliberately limited to upper-class residential areas, with new service mains following Jakarta's new "modern strip" of elevated highways and overpasses, hotels and high-rises, and the Asian Games complex. Network access did not extend to the "unmodern" spaces and populations thought to "lower the status of the nation"[49] (see map 4.1).

Although the map depicts Jakarta, it mirrors the distribution of water supply networks in many cities around the world. Networks are concentrated in key economic nodes (industrial, financial and banking, and elite

MAP 4.1. *Access to water supply in Jakarta (adapted from Bakker et al. 2008)*

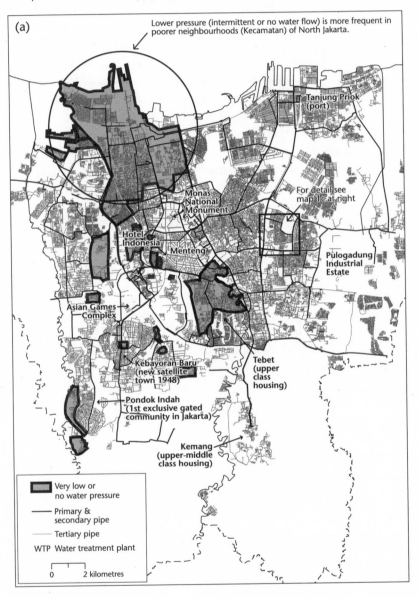

(b)   Major 'slum' areas, City of Jakarta (2005 data)

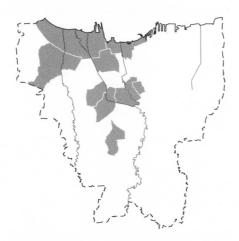

Slum households per Kecamatan > 15 %

(c)

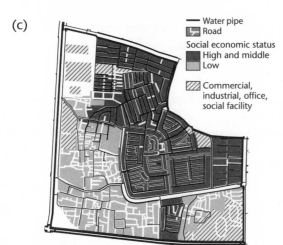

Water pipe
Road
Social economic status
High and middle
Low

Commercial,
industrial, office,
social facility

Low income areas have lower density of
water distribution pipes

0   200 metres

commercial) and higher-income residential zones. Most poor neighborhoods are excluded from network access due to the spatial distribution of water mains and service pipes. Lower or no network coverage is associated with higher rates of poverty. Moreover, areas of low or no pressure, as indicated on the map, tend to be concentrated in the poorest area of the city—here, in North Jakarta—the area of original colonial settlement and of the oldest and most poorly maintained infrastructure. Even in neighborhoods with water mains, access to networks varies at an extremely fine spatial resolution, often on a street-by-street basis (as illustrated by the inset map 4.1c). Pipes are densely and evenly distributed in the main thoroughfares and streets, where high- and middle-income households are found; network access is rare in the back alleys and side streets where low-income households predominate. Exclusion of the poor from access to water is literally built into the city's water infrastructure networks.

PRODUCING SCARCITY

What, then, explains the failure of the government to extend networks to poor households in the contemporary period? One important dimension of the answer to this question is the culture of governance in Jakarta. After independence, urban services provision and the urban environment were given relatively low priority, despite sporadic national government-led development plans to accelerate service delivery.[50] Official concern about the fact that the majority of the urban population could not afford to connect to the network was generally low.[51]

At the end of the first decade of independence, piped water supply was only available to one in ten Jakarta households. Network expansion barely kept pace with the city's exponential population growth after independence. By the 1990s, only 10 percent of lower-income residents were directly connected, although over 60 percent of the upper-income population had direct access to network water supply. Large tracts of the poorest areas of the city have remained without access to water supply networks altogether.[52] In a striking parallel to colonial disparities in water access, the majority of kampung residents continued to be served with public hydrants or by water vendors; they consumed less than 10 percent of the water distributed by PAM Jaya, compared to the approximately 60 percent consumed by the households with individual household connections.[53]

Why would this be the case? In the 1970s, policies that prohibited the provision of piped water to the rapidly growing informal settlements around the city were rationalized as a means of discouraging rural migrants, who were blamed for overtaxing the city's public services.[54] However, antimigration policies were not the sole factor. Network extension to poorer neighborhoods did not greatly improve after the "closed city" policy was relaxed in 1976.[55] Indeed, the first project focused on urban water supply to the poor occurred only in the 1990s.[56] Part of the explanation stems from the fact that, despite the failure of planning controls to stem urban sprawl, network expansion was limited to target zones in official planning, thereby excluding the irrigated farmland periphery where unregulated urban development was occurring.[57] The quiescence of consumers and complacence of government workers in the absence of popular politics during decades of dictatorship was equally, if not more, important.[58]

This lack of official concern was compounded by deep-rooted distinctions made by water supply workers between different parts of the city: between "elite" ("well-established" and "formal") service zones and the kampung neighborhoods that were literally "off-limits," ungovernable, and therefore unserviceable.[59] Stories abound of local mafia groups controlling public standpipes, actively discouraging (sometimes violently) attempts to install distribution networks or substitute household water connections for water hydrants.[60] Others refer to the difficulties stemming from the perceived low social capital of poor households; staff might be reluctant to deal with the poor because of illiteracy or lack of ability to correctly fill out forms, for example.

These views were both reflected and reinforced by PAM Jaya policies that required full cost recovery from customers after the early 1970s and limited network extension to neighborhoods where costs could reliably be recovered.[61] The tariff policy similarly discouraged the utility from connecting poor households: public hydrants were charged higher volumetric tariffs than individual households, creating a counterintuitive cross-subsidy from poor to middle- and upper-class customers, and implying that water revenues would decrease if large numbers of poor households were to be connected.[62]

Indeed, the water utility did not even have targets for supply coverage (in terms of the percentage of the city's total population) until the 1970s, and did not incorporate a poverty-reduction strategy in network-extension policy until the late 1980s.[63] So unlikely was network extension in the minds of water managers and policymakers that reported coverage figures for the city were usually stated as proportions of the total population in the target "served area," excluding kampungs altogether. Poor households were both figuratively and literally off the map; their residents were not considered to be deserving of, or able to connect to, the water supply system. As in many parts of the world, the revenues obtained from water bills represented a significant source of income for Jakarta's government, which was content to maintain the inequitable tariff structure.[64]

It should be emphasized that the spatial heterogeneity of the network as presented here was not solely due to a lack of water supply or of water resources availability during the colonial and much of the postindependence period. Indeed, the colonial water-supply system could theoretically have satisfied Jakarta's entire population (of approximately 800,000) until the end of the 1940s, assuming a per capita water demand of fifty liters per day and reasonable (e.g., 30 percent) leakage rates.[65]

This surplus of potential over actual supply persisted after independence, when production capacity consistently outstripped distribution capacity from the 1960s onward.[66] In particular, from 1965 to the late 1980s, the rehabilitation of existing infrastructure and the construction of an additional large-scale water-treatment plant increased water supply production capacity threefold. But by 1990 the water supply system still delivered only 40 percent of the potential volumes of treated water, and the provision of piped water supply still extended to less than one-quarter of the city's population.[67] Yet despite excess production capacity and a rapidly growing population, the city's water utility embarked only on limited extensions of the distribution network and increased production capacity largely in support of industrial users in the 1970s and 1980s.[68] In other words, a shortage of water resources was not the primary or sole reason for the failure to extend water supply network connections to poor households. Scarcity of potable water was indeed experienced in the city, but this was "second-order" or "social resource scarcity" rather than "first-order" or "natural scarcity."[69]

The preceding analysis suggests that a shortage of water resources was not the reason for the failure to extend water supply network connections to poor households in Jakarta. Throughout its history, the water supply system produced more water than it distributed. Even into the 1990s, after decades of investment designed to extend the network, the water supply system still delivered only 40 percent of the potential volume of treated water.

Development analysts sometimes call this phenomenon "elite capture," defined as "the process whereby elites 'skim' development resources intended for legitimate development ends and define policies in a way that protects their own interests."[70] This phenomenon has sparked significant debate within broader discussions over equity, growth, and development.[71] In turn, this has generated a wave of "pro-poor" development policies intended to address equity issues. But the assumptions implicit in these policies are, as the next section describes, at times far from accurate.

ARE THE POOR REALLY THIRSTY?

As in the colonial period, the contemporary behavior of poor households in Jakarta presents observers with an apparent paradox. Informal water services thrive in neighborhoods where network connections are available, with households relying on water vendors even when they have the option of in-house connections with the municipal water utility.[72] Although they are depicted as desperately needy of water in policy documents, members of poor households (when asked) rarely mention lack of water as their primary concern.[73] These households are dealing with many difficult challenges: often lacking secure income, the poor in Jakarta are faced with long commutes, polluted living environments, rapacious landlords, and government officials indifferent to many of their concerns, such as access to doctors and schooling for their children.[74]

But why would water figure low on the list of priorities of urban poor households? And why would water not become the target of popular political movements agitating for basic services—as is the case in many other parts of the world?

Easy access to groundwater everywhere in the city is one explanation. Shallow, hand-dug wells are a free alternative to buying water from the network or vendors. But the water is too polluted to drink, and the wells run dry in the summer months. Illegal connections are another explanation, but Jakarta is not as riddled with self-built "spaghetti" networks (so-called because of the appearance of the use of multiple flexible PVC pipes) as other cities.[75]

Decades of suppression of protest of any kind by the dictatorship that ruled Indonesia until 1998 are another explanation. But abrupt rises in gas and food prices at the end of Suharto's regime provoked massive demonstrations, and Jakarta's population still mobilizes regularly in political rallies. Water, in this case, is not the incendiary political issue that development experts and international campaigners imagine it to be; the efforts of Indonesian NGOs to mobilize large-scale campaigns around water supply largely failed to arouse public interest in the 1990s and early 2000s.

Another explanation—with a postcolonial inflection—relates to ways in which poor households perceive water quality. Kampung residents' perceptions of clean and safe water are associated with an absence of odor (including residual chlorine) and sediment. Even those with access to the city's piped network water supply system will aerate their drinking water in open buckets prior to consumption to dissipate the chlorine smell.[76] Some distrust the piped water system, preferring to buy water from a deep well bored by a wealthier neighbor. This is not without reason; the sole medical study to date that has compared the quality of tap and groundwater in Jakarta found that *E. coli* contamination is lower in shallow groundwater than in the piped water network (which is subject to frequent pressure reductions and leaks that compromise water quality).[77] Others prefer to purchase from water vendors, who check water quality and may strain the water or let it settle before delivery. And some kampung residents still prefer to wash their clothes, bathe, and defecate at the river or canal, despite having a well; running water is perceived to be cleaner than "still" (ground- or tap) water.

Policymakers have seized on these practices to justify hygiene campaigns, criticizing the "mystical" beliefs of the poor about the relationship between odor, water, illness, and disease transmission. The fact that the idea of cleanliness among kampung communities is "based on culture and tradition," rather than bacteriological criteria, reinforces the idea that people, along with pipes, are in need of development.[78] In a curious echo of colonial commentators, development policy writers depict the preferences of poor households for local water sources, "traditional" assessments of water quality, and the use of public spaces for domestic functions as evidence of the need for cultural and value change.[79]

The need for policy change is thus conveniently overlooked. A significant proportion of the city's population—the majority, in poorer neighborhoods—lives in rental or temporary (often self-built) accommodation without the legal residency permits and official (registered) land permits that are necessary to obtain a piped water connection. Deep wells are too expensive to build and maintain for those without permanent tenure. Moreover, landlords are often unwilling to connect rental properties to the water supply network because of concerns about infrastructure cost and maintenance; similarly, tenants are unwilling to connect, because their investment would constitute an upgrade to the landlord's property.

High connection fees and expensive fixed charges (such as the monthly "meter fee"), which may be five to ten times as high as the cost of water, mean that the effective cost per unit volume of tap water is higher than that of vendor water for the poorest consumers.[80] And that is even before adding in the cost of the water tank required because of the intermittent nature of water supply through the pipes (with cutoffs of several hours occurring daily in some areas), assuming that poor households have access to a roof-top on which to install the tank. Connection fees also vary depending on distance from the network; poor households are more likely to be located in areas of lower network density and thus to pay higher fees for connecting.

The water network may also simply be too inconvenient for poor house-holds. Low pressure in poorer areas of the city implies that water flow is intermittent, often resulting in greater infiltration of pollutants into the network and thereby reducing water quality. Low pressure also means that households prefer to have a backup source, usually a shallow well. During the dry season, water vendors represent a more secure source of water for households than the network. Long waiting times at water utility offices to pay bills and deal with meter misreadings, and bribes demanded by con-tractors and water utility staff—negligible amounts for wealthy households, but significant for the poor—are difficult for poor households to negotiate. The water utility, depicted as the modern ideal in development policy docu-ments, compares poorly with the service provided by vendors, who visit the home to collect bills and allow flexibility in payment, sometimes even permitting households to buy water on credit.[81]

Development discourse also obscures the fact that the choice not to use the water supply network offers important, symbolic vehicles of resistance to poor households, often distrustful of state authority that all too often has enforced forms of development that undermine rather than improve their livelihoods.[82] Using the public fountains in the city's iconic Independence Square as a place for bathing and laundry, low-income residents have sup-planted the intended "spiritual symbolism" of this urban space with politi-cally symbolic calls for the government to meet immediate material needs.[83] Illegal connections (estimated at forty thousand) have provided another important means of resistance.[84] Over half of Jakarta's network water sup-ply is lost through "unaccounted-for water" (UFW), 80 percent of which is estimated to be from "administrative" rather than physical leakage.[85]

MARKET FAILURE: JAKARTA'S EXPERIMENT WITH PRIVATE SECTOR PROVISION

By the early 1990s, the persistent failure of government-run water-supply projects had become a mantra of development-agency policy. This, of course, neatly coincided with the rise to prominence of the Washington Consensus. In Jakarta, as elsewhere, private sector "participation" in man-aging the water supply system was sought. Private sector operators would, it was hoped, be able to improve cost recovery in order to facilitate repayment of the substantial debts owed to international financial institutions by the water utility and enable extension of the network to Jakarta's kampungs. Like many capital cities, Jakarta was targeted by the multinational water companies eager to extend their markets. Jakarta was particularly appealing to private water companies because of its relatively low rate of household connections, its large middle class, and the apparent economic and political security offered by a long-standing dictatorship.

Discussions with foreign firms regarding privatization began in the mid-1990s. After protracted negotiations, "cooperation agreements" for the management and expansion of Jakarta's water supply system were awarded in late 1997 to two of the largest water-services companies in the world: the British firm Thames Water International and the French firm Ondeo (Suez-Lyonnaise des Eaux). The twenty-five-year contracts divided the city into two halves run separately by the local subsidiaries of these two operators, Thames PAM Jaya (TPJ) and Ondeo's subsidiary, Palyja. Ambitious targets were set: the private companies committed to supplying potable water by 2007 and to reaching universal coverage by 2023. The contracts provided substantial risk protection for the private companies: each private partner was reimbursed by the municipality on the basis of volume of water supplied and billed, not on the basis of actual revenues (removing the risk of low cost recovery). An additional safeguard was built into the payment mechanism: an indexation formula, linked to the Indonesian rupiah–U.S. dollar exchange rate and the (Indonesian) inflation rate, was built into the "water charge" formula used to determine payments made to the private operators. Cost recovery and currency risks, in short, were to be borne by the local government.

The political and economic turmoil that unfolded in Indonesia in 1998 stymied these strategies. Riots, the resignation of Suharto, and the abrupt and dramatic devaluation of the Indonesian rupiah threw the country into a period of chaos.[86] Now free to protest, water union members barricaded company offices and picketed headquarters, criticizing the corrupt contracts. The two companies had put forward unsolicited proposals directly to the government and were then partnered with local companies associated with close contacts of the president.[87] After a tense interlude in which senior expatriate managers of the private concessionaires fled the country, local managers canceled the contracts, and senior British and French executives and diplomats pressured the federal government to have the contracts reinstated, the private concessionaires resumed operations (having discreetly abandoned their Indonesian partners, now tainted by their association with the ex-president).[88]

Confronted with public protest over rising prices of staple food items and gasoline, the municipal government refused to raise water tariffs to compensate for the devaluation of the rupiah. Receipts in dollar terms plummeted, and PAM Jaya (and its government owner) thus became increasingly indebted to the private companies, with the cumulative deficit reaching almost US$100 million by late 2003.[89] The prospect of slow repayment of the still-increasing debt provoked a renegotiation of the original agreements, which were scaled back into management (rather than concession) contracts with a guaranteed internal rate of return of 22 percent. Performance targets were also dramatically scaled back (table 4.1); most notably, the commitment to provide potable water supply at the point of consumption was dropped, and targets for network extension were reduced.

Despite this, the companies had still failed to reach their original targets by 2004, although two key performance measures did show steady improvement by 2008 (service coverage ratio and numbers of connections).[90] And although some new connections were installed, these were disproportionately weighed toward middle- and upper-income households (figure 4.1); over the period 1998–2004, only 25 percent of new connections were targeted in the two lowest tariff bands (public hydrants, intended to serve those without household connections; and "poor" and "very poor" households). In other words, three-quarters of new connections were for middle-income and upper-income households, government enterprises, and commercial enterprises (table 4.2).[91] Given that the private companies' revenues were linked to volumes of water delivered into the water supply system (rather than revenues and tariffs), this bias toward wealthier consumers

TABLE 4.1. *Original versus Renegotiated Technical Targets, Jakarta (1998–2008)*

	Number of connections		Service coverage ratio (%)	
	Thames PAM Jaya	Palyja	Thames PAM Jaya	Palyja
Baseline (before privatization)	231,607	176,980	52.00	38.00
Original targets (1997)	361,607	395,522	70.00	70.00
Revised targets (2002)	335,413	301,048	62.00	45.00
Realization (2002)	336,550	312,879	62.17	44.17
Revised targets (2008)	403,030	391.980	75.50	61.00
Realization (2008)	379,480	398,507	65.28	61.58

Data source: PAM Jaya 2004; JWSRB 2008; TPJ and Palyja annual reports; Shofiani 2003.

4.1. *Customers per tariff band in Jakarta, 2003 (adapted from Bakker 2007b; data drawn from PAM Jaya 2004)*

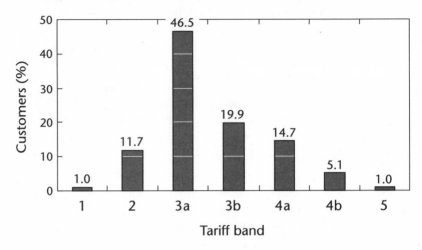

might seem surprising. There was thus no apparent direct disincentive to the private concessionaires to connect low-income households. Why, then, were customers in the lowest tariff bands less likely to be connected? In part, because the price of the water sold to the poorest customers was set well below production cost; connecting more poor households implies a reduction in revenue, in turn reducing the municipality's ability to repay the debt shortfall owed to the private operators.[92]

This failure to connect poor households undermined the central justification for involving the private companies: the belief that this would lead to a significant extension of the network into the kampungs. But "pro-poor" activities on the part of the private concessionaires were constrained by the inability of poor households to pay the high connection fees and were limited to a few showcase projects.[93] One of the two private concessionaires, for example, began allowing households to pay connection fees in monthly installments over a twelve-month period rather than demanding full payment up front. But other significant issues, including low water pressure in poor neighborhoods, the overall burden of connection fees and transaction costs, the absence of pro-poor billing practices, the inability-to-pay of poor households (particularly those with fluctuating incomes), and poor water quality, were not addressed by the private concessionaires.

The difficulty of extending water supply to low-income households is demonstrated by the outcomes of an ambitious World Bank project, initiated in 2006, which promised twenty thousand new, directly subsidized

TABLE 4.2. *New Connections, East Jakarta 1998–2004, by Tariff Band*

Tariff band	Tariff group	Number of new connections	Increase (%)
1	Social institutions and public hydrants	1,101	1
2	Public hospitals, poor and very poor households	21,898	24
3a	Middle-income households and small-scale businesses	51,847	58
3b	Upper-middle-income households and government offices	11,150	12
4a	Large hotels, high-rise buildings, banks, and factories	2,323	3
4b	Harbor/port	1,849	2
Total		90,168	100

Data source: Thames Pam Jaya, personal communication (May 2005).

connections to poor households. The project encountered a stumbling block early on: most households considered to be sufficiently poor to receive a subsidized connection do not have legal land-tenure status, so cannot receive a subsidy. Despite lobbying of government by the private sector operators to change this tenure policy, the original target was scaled back to fewer than four thousand households.[94] Even direct subsidization of the private sector operators by the World Bank had failed to enable water supply network access to be extended to the city's poorest households.[95]

Meanwhile, one of the two private operators (Palyja) reported an increase in the number of low-income households served: from 11,659 in 1998 to 69,581 households in 2007, a total increase of 57,922 households (PALYJA 2008). This figure should be contrasted however, with a total increase of 103,154 for middle- and upper-class households over the same period (PALYJA 2008).[96] Despite being a small proportion of the total households, wealthier households continue to have the majority of in-home connections, and benefit disproportionately from new connections. Moreover, these figures pertain to only the western half of the city (and are likely to be much worse in the eastern, poorer half).

By 2009, a decade after the start of the contract, government interest in rolling out connections to lower-income households remained a low priority, for two reasons. First, political pressure to keep tariffs low limited the potential for cross-subsidization. Second, a new emphasis on network connections for commercial and high-end properties using deep groundwater (largely in the central business district) meant that new investments would largely be focused on increasing supply capacity via network rehabilitation, rather than expanding the network into unserviced areas or improving supply in low-income areas.

Although the private sector contract had been negotiated and signed during the reign of the Suharto government, by the time the companies returned to operations following events of 1998, the civil-society groups that had flourished following Suharto's fall from power had launched a concerted campaign against water privatization. Consumers and environmental groups joined forces to criticize the record of the private companies. Although court cases against the municipal and federal governments failed, the public attention drawn to water issues in the capital increased political pressure on Jakarta's governor to rein in tariff increases and to push the companies to improve their lagging performance. Relations grew progressively more strained; by 2006, one of the private companies had withdrawn, and the other had substantially modified its role in running the water supply system, thereby reducing its risks.[97]

CONCLUSIONS: THE "POLITICS OF THE GOVERNED"

Our analysis of the history of Jakarta's water supply network suggests that the efforts of both governments and the private sector with respect to the urban poor are ambivalent; and the actions of both governments and private companies undermine, at times, the ability of communities to engage in collective action to solve the problems posed by urban water provision. But the story is more complex than a simple public-private binary implies, because it is one in which poor households also play a role in finding alternative solutions, some of which are neither public nor private, and which are sometimes expressive of resistance to the remedies prescribed by development experts. It is also a story in which market failure, state failure, and governance failure combine to fragment access to urban water supply.

To illustrate this point, remember the dilemma first discussed in chapter 1: governments often seem incapable of providing services on which their legitimacy depends, and the elite often determinedly exclude the poor. One response, as in India (Chatterjee's focus), is that these claims are continuously negotiated between governmental obligations to look after the underprivileged, on the one hand, and political expediency on the other. In other countries, such as Indonesia, the predominant response is a retreat of the elite—a "variant of the colonial strategy of indirect rule," in which the modernization project is suspended, and the protected zones of bourgeois civil society are literally walled off—in Jakarta's case, through gated housing complexes, private malls, and office precincts, connected by a network of toll roads.[98] The project of modernization, which is, in part, a process of converting subaltern subjects into citizens (and mutually transforming these groups in the process), comes to a halt; or, more accurately, is transfigured as a dual process of inclusion and exclusion (evocatively termed "global apartheid" by Patrick Bond).[99] The resulting social and spatial fractures are particularly evident in capital cities like Jakarta, because of their urban primacy, and their role as a political symbol and node of economic activity. Here, as elsewhere, public resources are spent on privileged consumption by a small minority of urban residents, and only these citizens are considered to be political constituents of society with full entitlements to state services.

To expand on Chatterjee's arguments, we would simply add that these issues are particularly fraught when local, common-pool resources are at stake. It is here that market failure, state failure, and governance failure are likely to emerge and reinforce one another. Unequal access to water supply in Jakarta, as in many cities around the world, is thus not simply the outcome of neoliberalism or vulture capitalism (the "market failure" argument), or of incompetent or self-interested governments (the "state failure" argument). Unequal access (scarcity for some, but not for others) was literally hardwired into the city's network by colonial powers and sustained in the postcolonial period by governments of both left and right political persuasions, as well as the private sector operators brought in to manage the city's water supply system in the 1990s. The distinction between citizens and populations is actually built into the city, through the interrelated production of subjectivities (Europeans versus natives), spaces (the city and the kampung), and infrastructure (the network versus the archipelago). This is simultaneously a cultural as well as a political and economic process: a lack of financial means is certainly an important factor, but so too is the Indonesian experience of modernity and its articulation of the tastes, desires, and expectations of its elites, modulated by its postcolonial history and nascent nationalism, in turn mediated by development programs and technologies originating from the West. As a result, in both the colonial and contemporary periods, residents whose domestic water practices do not demonstrate a familiarity with scientific rationalities, modern concern for bodily health, or an appropriately economical use of water have been marginalized both materially and discursively. Access to a modern piped water supply became contingent on identity (as a "European" colonial subject, or as a "modern" citizen).

This provides an alternative viewpoint from which to consider water supply access in a city like Jakarta. Both public and private operators of the water supply system are subject to constraints that relate to questions of culture (such as the colonial discourse of the "ungovernable" kampungs) as well as economics, and to norms of urban governance as well as utility performance. The seemingly paradoxical resistance to connecting the urban poor living in illegal settlements only becomes legible if this debate is framed in terms of Jakarta's culture of urban governance and the Indonesian experience of modernity in which this culture of governance is embedded. The "governance failures" evident in Jakarta encompass, on the one hand, practical issues (such as targeting, internal incentives, and the capacity to provide services in a manner amenable to the needs of the poor) and issues of power and politics (such as elite capture of urban infrastructure) on the other.

What, then, should our response be to these inequalities (which remain sufficiently dramatic as to be self-evident, although their historical evolution may be a matter of dispute)?

One response, from those concerned about the persistent lack of access to water in cities like Jakarta, has been to expand the claims of subjects and to pursue the project of converting subjects into citizens with full rights. The international campaign for the human right to water pursued by nongovernmental organizations over the past decade is one such example. Yet as we have suggested in this chapter, when confronted with claims for rights in the absence of material means to provide those rights, governments may retreat, obfuscate, or respond with repressive and exclusionary forms of social control. This interrelationship between processes of exclusion and modernization suggests that questions of community management should also be the focus of our attention. These are two of the key issues addressed in part 2 of this book.

BEYOND PRIVATIZATION
DEBATING ALTERNATIVES

5

PROTESTING PRIVATIZATION

TRANSNATIONAL STRUGGLES OVER THE

HUMAN RIGHT TO WATER

On a rainy Friday in 2003, the world's water and environment ministers met in Kyoto to discuss the global water crisis. While ministers met behind closed doors, participants in the concurrent World Water Forum were presented with alarming statistics: water scarcity had been growing in many regions, and over 20 percent of the world's population was without access to sufficient supplies of potable water for basic daily needs. Conference organizers had drafted an Inter-Ministerial Declaration, which reflected the view that the best response to increasing scarcity was the commercialization of water.

Accordingly, private water companies had been invited to meet with government delegations, international financial institutions, and bilateral aid agencies to develop solutions to the world's water problems. Yet many of the governments represented at the conference had themselves been accused of irresponsible water management by their citizens.

Representatives of private water companies were similarly targeted by activists: corporations such as Enron came under attack by an international alliance of anti-dam campaigners, environmentalists, public sector unions, international "bank-watcher" think tanks, alter-globalization activists, indigenous peoples' organizations, and civil-society groups. These self-proclaimed "water warriors" protested both inside and outside the forum, criticizing the forum's co-organizers (the Global Water Partnership and the World Water Council) for their close ties to private water companies and international financial institutions.

Activists' protests culminated with the disruption of a planned highlight of the forum, a plenary session, chaired by a former head of the International Monetary Fund, promoting active government support for increased private sector involvement in the water sector in lower- and middle-income countries.[1] Chanting "water is life," activists stormed the stage and demanded the withdrawal of the private sector, a return to local "water democracy," a rejection of large dams as socioeconomically and environmentally unsound, *and recognition of water as a human right. Yet activists' calls fell largely on deaf ears. Ministerial delegates reached consensus, including, controversially, support for private sector financing and new mechanisms for private sector involvement in water supply management. Notably, as with previous World Water Forums, the joint ministerial statement conspicuously failed to refer to water as a human right.*

The human right to drinking water is perhaps one of the most disputed issues within development debates at the start of the twenty-first century. For many, it comes as a surprise to learn that a right to water is not specified in any of the primary United Nations conventions on human rights.[2] Some governments (including the Netherlands and the United Kingdom) express their public commitment for a human right to water, and a few (such as France, South Africa, and Uruguay) have embedded the right in legislation.[3] But they are the exception, and some governments (such as Canada) have rejected calls for the creation of a human right to water.[4]

This gap in human-rights legislation was, for many decades, largely overlooked. Usually this was because an "essential services" or "basic needs" approach to water supply was assumed to be sufficient. But the emergence of an urban water crisis and the debate over water privatization, in the context of a more generalized surge of demands for a "human rights–based" approach to development in the 1990s, triggered what has now turned into a full-blown global campaign for a human right to water.[5] Many antiprivatization activists have focused their activities on this campaign, framing "water as a human right" as a means of ethically refuting and practically constraining privatization.

The debate over the human right to water raises several questions. Is water a human right? If so, is private sector provision incompatible with the human right to water? And what is the relationship between property-rights regimes, the human right to water, and privatization? This chapter explores these questions, documenting the different constructions of rights adopted by pro- and antiprivatization advocates, questioning the utility of the language of "human rights," and interrogating the accuracy of the (often unquestioned) dichotomies—rights/commodities, public/private, citizen/customer—deployed by both sides of the debate. (Water rights, as a form of property rights, are a distinct although clearly related issue and are considered in chapter 6).

In the first part of this chapter, I briefly describe some of the key arguments of the antiprivatization campaigns, networks, and coalitions that have emerged since the early 1990s (such as the "Blue Planet Project," the

"Reclaiming Public Water" network, "Octubre Azul/Blue October," the "Right to Water" coalition, and alternative global water forums). In the second half of the chapter, I focus on debates over the human right to water. I take issue with the ways in which antiprivatization campaigns have mobilized calls for the human right to water, and examine the limitations to, and potential tactical advantages of, a human right to drinking water.

My core argument is that the human right to drinking water is potentially useful in aspirational terms, and also in clarifying the roles and responsibilities of public and private actors. But a human right to water is unlikely, in isolation, to create conditions for achieving the goal of "water for all," in part because it rests on an overly restrictive framework through which to deal with broader issues of community water rights and ecological sustainability. Nonetheless, the new political terrain opened up by "rights talk" offers progressive possibilities for collective action, which I explore in the closing sections of the chapter.

PROTESTING PRIVATIZATION

Protests against water privatization are remarkable for the diversity of the coalitions they engender. It is not unusual to find environmental, women's, antipoverty, and religious groups working together with trade unions in nascent red-green alliances from local to global levels. Often, privatization protests are linked to, and share much in common with, alter-globalization[6] struggles more generally (particularly those calling for reform of international financial institutions, such as the World Bank, and trade agreements and organizations, such as the GATS and WTO).

This diversity means that it is difficult, if not impossible, to accurately characterize the antiprivatization movement. Nonetheless, some common threads in the messages of privatization protesters can be identified. For many antiprivatization activists, water privatization is an act of what David Harvey terms "accumulation by dispossession"[7]—the appropriation, for profit, by the private sector of both the natural environmental "commons" and the public goods created and heavily subsidized by the state. This act of dispossession is emblematic of "globalization from above," with the negative consequences that this process is assumed to entail.[8] According to its opponents, the involvement of private companies invariably introduces a pernicious logic of the market into water management, which is incompatible with guaranteeing citizens' basic right to water. Private companies, answerable to shareholders and with the overriding goal of profit, will manage water supply less sustainably than public sector counterparts. Opponents of privatization point to successful examples of public water systems and to research suggesting that private sector alternatives are *not* necessarily more efficient, and are potentially more expensive for users, than well-managed public sector systems (as discussed in chapters 2 and 3). They assert the importance of democratic accountability to citizens and the insufficiency of corporate accountability to shareholders as a substitute. In voicing opposition to privatization, activists point to the fierce debate provoked by the increasing involvement of private, for-profit, multinational water corporations in running networked water-supply systems around the world.[9] Their position is that water is a nonsubstitutable resource essential for life, and they call for water supply to be recognized as a human right, placing an onus on states to provide water to all and to prevent private sector involvement.[10]

Accordingly, most antiprivatization activists reject the conceptual bases on which proponents of privatization base their claims. They do not agree that water should be priced at its full cost; nor do they agree that this will lead to more equitable outcomes. They reject the claim that private management, on a for-profit basis, will improve performance; and they equally reject the claim that accountability to customers and shareholders is more direct and effective than the attenuated political accountability exercised by citizens via political representatives.[11]

Underlying this stance is the assumption that public goods are inviolably collective, and that water, as the biophysical and spiritual basis of life, is the public good par excellence. Defending the sphere of public goods—a reserve of natural and social capital under threat from capitalism—is the broader task that antiprivatization activists see themselves as carrying out. The notion that water has an economic *value* engenders some agreement; but the notion that water is therefore an economic *good* is more controversial.

These campaigners frequently adopt a definition of equity distinct from that of privatization advocates. For the latter, equity is usually defined in economic terms; in pricing, for example, full-cost pricing is predicated on a concept of economic equity, defined according to the "benefit principle" (users should pay the full costs they impose on the water supply system). For the antiprivatization side, in contrast, equity in pricing is defined in social terms; "social equity" or the "ability-to-pay" principle implies that users should pay for water in line with their ability to do so, suggesting the need for cross-subsidies between sectors and classes of consumers. Moreover, these incompatible definitions of equity—and the conflict, almost a truism of the sustainable-development literature, between efficiency and fairness that they invoke—are sometimes rejected by antiprivatization protesters.

Rather, challenging the false paradox between equity and efficiency, protesters reason in favor of prioritizing concepts of redistribution (rather than
production) and of altering technical and financial aspects of water distribution systems accordingly. Subsidies for water-efficient appliances and
in-house connections, for example, are sometimes suggested as a means to
combat "water poverty" through simultaneously promoting water conservation and access.[12] Others who reject privatization nonetheless admit to
the importance of water pricing as a mechanism for conservation.

This ambivalence is not exclusive to activists. Take, for example, the language of the EU Water Framework Directive, arguably the most sophisticated piece of water legislation in the past decade. The directive's first article
(subjected to protracted wrangling as it was being drafted) reads: "Water is
not a commercial product *like any other* but, rather, a *heritage* which must
be *protected, defended, and treated as such*" (emphasis added).[13] However,
much of the directive stresses the need (and creates obligations for member
states) to value water's economic uses and to reorient management according to cost-benefit analyses. Similarly, the UN Committee on Economic, Social, and Cultural Rights recognized this ambivalence when it defined water
as a social, economic, and cultural good as well as a commodity.[14] And the
Dublin-Rio Principles (perhaps the most influential set of water policy principles over the past two decades) assert that water has an economic value
and state that "it is vital to recognize first the basic right of all human beings to have access to clean water and sanitation at an affordable price."

Should we accept that water is a commodity? A compromise position
might be to argue that while water has economic value, it cannot simply
be defined as a commodity, for both ethical and technical reasons. A more
radical critique offered by antiprivatization scholars and activists frames
privatization as part of a broad-based intensification of the "accumulation"
of nature under capitalism, in which nonmonetary values and community
management are displaced by economic values and private sector–controlled
(or market-mimicking) ownership, management, and allocation. Market-
mimicking techniques, from this perspective, are not the benign, neutral
management techniques presented in most of the contemporary water management literature. Resistance to privatization is thus, in many instances,
also linked to resistance to markets (and capitalism) in a broader sense. As
such, antiprivatization campaigns and protests are emblematic of Polanyi's
"double movement," in which the attempt to dis-embed the economy from
society (and from social constraints on markets) provokes a corresponding
attempt to re-embed private sector activity within society, through reimpos-
ing social constraints on markets and private firms.

Where have these protests taken place, and have they been successful? Protests against water privatization have been fairly widespread, as evidenced in table 5.1. However, not all water privatization initiatives have sparked large-scale protest. Privatization in some countries (such as China) has provoked little organized resistance.[15]

TABLE 5.1. *Public Protests against Private Sector Water Supply (selected examples)*

Country	Place	Date
Argentina	Buenos Aires	Mid- to late 1980s
Argentina	Buenos Aires province	2002
Argentina	Córdoba (province)	2006
Argentina	Tucumán and other parts of Argentina	1993–99
Bolivia	Cochabamba	2000
Bolivia	El Alto and La Paz	2005
Brazil	national	2002
Brazil	Rio de Janeiro	1999
Canada	Montreal	1999
Canada	Sackville	2007
Canada	Vancouver	2001
Colombia	national	2007
Ecuador	Quito	2007
El Salvador	Santa Eduviges	2006
El Salvador	San Salvador	2006
England	national	1985–89
France	Grenoble	2001
Germany	Munich	1998
Germany	Potsdam	2000
Ghana	Accra	2002
Honduras	national	1995
Hungary	Debrecen	1995
India	New Delhi	2003–2005
Indonesia	Jakarta	2003
Kenya	Nairobi	2007
Malaysia	Kuala Lumpur	2005
Mali	Bamako	2007
Mauritius	national	2000
Mexico	Mexico City	2006
Morocco	Rabat	2006
Nepal	Kathmandu	2006
Panama	national	1999

(continued)

TABLE 5.1. *(Continued)*

Country	Place	Date
Paraguay	national	2002
Peru	national	2004
Poland	Lodz	1994
Poland	Poznan	2002
South Africa	Alexandra	2002
South Africa	Gauteng, Mpumalanga, and Eastern Cape provinces	1997–98
South Africa	Johannesburg and townships	2001–present
South Africa	Nkonkobe	2002
Sri Lanka	national	ongoing
Sweden	Malmo	1995
Tanzania	Dar es Salaam	2005
Thailand	national	2002
Trinidad	national	1999
Ukraine	Odessa	2000
Uruguay	national	2002
Uruguay	Montevideo	2003
USA	Atlanta, Georgia	2003
USA	Birmingham, Alabama	2000
USA	Mecosta County, Wisconsin	2001
USA	San Francisco, California	2003
USA	Santa Margarita Water District, California	1995
USA	Stockton, California	2002
USA	Washington, DC	1996
USA	Wisconsin Dells, Wisconsin	2000

The list of countries provided in table 5.1 indicates that protests against water privatization have occurred in countries of all income levels. This suggests that relative impacts on income distribution, rather than absolute impacts on wealth, are critical factors in protests.

What are these protests about? Although the range of grievances varies, concerns over the impacts of privatization on equity (fairness in pricing and access to services), employment, and working conditions predominate, as suggested by the higher representation of consumer groups and unions. Environmental concerns and nationalist rejection of "foreign" ownership and management of water supply utilities have been less frequently represented than the concerns of consumers and unions, although in many instances a nascent "red-green" alliance has cohered around the antiprivatization agenda (as in Cochabamba, Jakarta, and Johannesburg). Unsurprisingly, antiprivatization protests are often articulated with, and amplified by, more generalized political-economic protest. Argentina and Bolivia, for example, experienced significant public protests on a range of economic issues from the late 1990s onward; water privatization was linked with broader concerns about macroeconomic management, the behavior of political classes, and the appropriate role of the state (and foreign capital) in the national economy.[16] As described in chapter 3, the concentration of multinational water companies during the wave of privatizations in the 1990s was twofold: urbanized areas, where economies of scale increase profit potential; and middle-income countries in east Asia and Latin America, where ability-to-pay was presumably higher. Social movements were particularly vociferous in Latin America, long characterized by strong traditions of economic nationalism and anti-imperialism.

In adopting these strategies, how successful have antiprivatization activists been? Cancellations of high-profile privatization projects are often held up as examples by antiprivatization advocates of the success of their campaigns. To a significant degree, large-scale protests, where they have occurred, were successful at encouraging the rejection or cancellation of private sector participation contracts.[17] In Cochabamba (Bolivia), Stockton (United States), and Tucumán (Argentina), for example, the consumer campaigns were successful; others (such as the Ghana campaign) were not.[18] But the analysis in chapter 3 suggests that these claims must be moderated by the recognition that many private sector contracts fail for technical reasons related to profitability and risks, rather than (or in addition to) protest.[19]

It also bears noting that these protests are often highly reactive, arising in response to either a proposed or implemented privatization contract. Few protests succeed in transcending the perceived threat at hand and translating themselves into sustained activity beyond the departure of the private company, although there are notable exceptions, such as the international campaigns for "remunicipalization" and "public-public partnerships."[20]

In this regard, the trajectory of the movement against water privatization differs from that of the anti-dam protests and campaigns. The latter, peaking in the 1980s and 1990s, were characterized by highly effective international advocacy networks, which were successful in incorporating anti-dam views into broader campaigns targeting international financial institutions such as the World Bank, and in forcing the dam industry to participate in an unprecedented inquiry into development practice, the World Commission on Dams.[21] In contrast, antiprivatization campaigns to date have been less tightly articulated with global campaigns (although water has featured relatively prominently in anti–GATS/WTO campaigns) and less effectively networked.[22] Moreover, these campaigns have not, to date, been as successful as the anti-dam movement in forcing changes in the policy status quo.[23] Where these changes have occurred, they have been as a result of internal dynamics within the water industry and international financial institutions, rather than as a result of external pressure.

This is the case, I would suggest, for two reasons. First, unlike the anti-dam movement, there is no single protagonist (the equivalent of the International Commission on Large Dams); although the World Water Council and Global Water Partnership are closely associated with private sector companies, they do not publicly represent proprivate interests in the same way that ICOLD does for large dams. Second, the case for alternatives to privatization is less clear than it is for large dams. Defending the status quo (government provision) is fraught with ambivalence, and the appeal and practicality of appropriate (alternative) technology for water supply systems (such as rooftop rain barrels and hand-dug wells) are not as great as they are with respect to small dams and alternative water-storage methods, both for public-health reasons and because of the symbolism of a household network connection as what I have elsewhere termed a material entitlement of citizenship.

An important range of "alternatives" to privatization has appeared in recent years. Table 5.2 is an attempt to structure our analytical frameworks for assessing these reforms. Why is such a typology necessary? As explored in chapter 1, reforms can be undertaken in distinct categories and are not necessarily concomitant. One may privatize without deregulating; deregulate without marketizing; commercialize without privatizing, and so on. To give a simple example: privatization of the water supply industry in England and Wales in 1989 did not entail marketization; that is, it did not entail the introduction of markets in water abstraction licenses. Here we can observe one of the main confusions in the literature: reforms to institutions, organizations, and governance are often subsumed under general labels (like "privatization" or "neoliberalization"), despite the fact that they often involve very different types of reforms, applied to different aspects of resource management. Another source of confusion arises when different types of reforms are assumed to be interchangeable, and when distinct terms (marketization, privatization) are assumed to be synonymous, when they are not.

How is such a typological exercise helpful in either analysis or activism? First, the failure to distinguish between categories of resource management, and between targets and types of reforms, obscures the specificity of the reform processes that are the object of analysis, and it limits our ability to

TABLE 5.2. *Water Supply Reforms and Alternatives*

Category	Target of reform	Type of reform	Alternative
Resource management institutions (laws, rules, norms)	Property rights	Privatization	Mutualization (recollectivization) of asset ownership (in Wales) (Bakker 2004) Communal water rights in village "commons" in India (Narain 2006)
	Regulatory frameworks	Deregulation	Creation of customer service boards with oversight and decision-making power Reregulation by consumer-controlled NGOs such as "Customer Councils" in England (Franceys 2006; Page and Bakker 2005)
Resource management organizations (collective decision-making and management entities)	Asset management	Private sector "partnerships"	Public-public partnerships (e.g., between Stockholm Vatten and water utilities in Latvia and Lithuania) (PSIRU 2006) Water cooperatives in Finland (Katko 2000)
	Infrastructure finance	Private capital flows	Public financing mechanisms (e.g., state revolving funds, municipal bonds) (WDM 2007) Microcredit Corporate social-responsibility criteria
	Organizational structure	Corporatization	Low-cost, community-owned infrastructure (e.g., Orangi Pilot Project, Pakistan) (Zaidi 2001)
Resource governance (decision-making process)	Resource allocation	Marketization	Sharing of irrigation water based on customary law (*usos y costumbres*) in the Andes (Trawick 2003)
	Performance incentives/ sanctions	Commercialization	"Customer Corporation" (with incentives structured toward maximization of customer satisfaction rather than profit or share-price maximization) (Kay 1996)
	User participation	Devolution/decentralization	Community watershed boards (Canada) (Alberta Environment 2003) Participatory budgeting (Porto Alegre, Brazil) (Balanyá et al. 2005)

Source: Adapted from Bakker 2007a

compare cases. For example, comparing the introduction of water rights for "raw" water (water in nature) in California to private sector participation in water supply management in New York is of limited interest, because two distinct processes are at work (marketization and private sector participation).[24] In contrast, comparing the introduction of water markets in Chile and California is worthwhile, because in both cases private property rights for water supply have been introduced via a process of marketization of water resource allocation.[25] In short, the typology presented in table 5.2 is useful because it enables us to correctly compare different types of market reforms and to more accurately characterize their goals and evaluate their outcomes.

This typology is also useful in addressing the widespread failure to adequately distinguish between different elements of neoliberal reform processes: this analytical sloppiness diminishes our ability to correctly characterize the aims and trajectories of neoliberal projects of resource management reform.[26] Commercialization, for example, often precedes privatization in the water supply sector, which is sometimes followed by attempts to commodify water. The biophysical properties of resources, together with local governance frameworks, strongly influence the types of neoliberal reforms that are likely to be introduced. Common-pool, mobile resources such as fisheries are more amenable to market-led allocation strategies, whereas natural monopolies such as water supply networks are more amenable to full privatization. In other words, in failing to exercise sufficient analytical precision, we are likely to misinterpret the reasons for, and incorrectly characterize, the pathway of specific neoliberal reforms.

This typology may also be useful in clarifying activist strategies and in structuring our analyses of activism and advocacy. For example, activists often describe water as a "commodity" in contrast to water as a "human right." But table 5.2 suggests that this is misleading, insofar as the term "commodity" refers to a property-rights regime applicable to resources, and human rights to a legal category applicable to individuals. The more appropriate, but less widely used, antonym of water as a commodity would be a water "common pool resource."

A notable absence in this typology is the human right to water. This might seem surprising, until we consider the claim that the notion of water as a human right is not necessarily incompatible with water privatization.

CAMPAIGNING FOR THE HUMAN RIGHT TO WATER

The international campaign for a human right to water has grown enormously over the past decade.[27] Beginning with a set of declarations by activists, a broad-based campaign has emerged bringing together development-focused aid watchdogs (such as the UK's World Development Movement), mainstream international organizations (such as the World Health Organization), human-rights organizations (notably Amnesty International), environmental groups (the Sierra Club, one of the largest U.S. activist groups), consumer groups (such as Ralph Nader's Public Citizen), and more radical antiprivatization and alter-globalization groups (such as South Africa's Anti-Privatization Forum and the Council of Canadians), as well as networks of influential international actors (such as Mikhail Gorbachev's Green Cross and Ricardo Petrella's Group of Lisbon).[28] Countering the claim that water is an economic good (and a human *need*), the Water Supply and Sanitation Collaborative Council's Vision 21,[29] the Cochabamba Declaration,[30] and the Group of Lisbon's Water Manifesto[31] on the right to water have insisted that water is a human right under international law.

Activists have also focused on country-specific campaigns for constitutional and legal amendments, notably in Uruguay, where a national campaign resulted in 2004 in a national referendum on the human right to water and a resulting constitutional amendment declaring water access a human right. Similar efforts on the right to water subsequently emerged elsewhere in Latin America (particularly Bolivia, Colombia, Ecuador, El Salvador, and Mexico).

As the campaign for the human right to water has gathered momentum, activists have gained support from mainstream international development agencies including the World Health Organization, the United Nations Development Program, and the United Nations General Assembly, whose president appointed a high-profile antiprivatization activist, Canadian Maude Barlow,[32] as senior advisor on water in late 2008.[33] This follows General Comment 15 issued by the UN Committee on Economic, Social, and Cultural Rights in 2002, asserting that every person had a right to "sufficient, safe, acceptable, physically accessible, and affordable water." [34] The committee declared that the right to water was implied in the International Covenant on Economic, Social, and Cultural Rights, and that states party to the covenant have the duty to realize, without discrimination, the right to water.[35] Importantly, the committee argued that realization of the right to water is linked to the realization of many other rights, including the rights to food, health, adequate housing, the right to gain a living by work, and the right to take part in cultural life.[36] And in 2008, the UN Human Rights Council adopted a resolution reconfirming the obligations of governments to ensure access to safe drinking water and sanitation, and establishing an independent expert on related human-rights obligations.[37] This means that the UN human rights system now has a separate mechanism exclusively dedicated to issues related to the right to water and sanitation.

These developments are celebrated by advocates, who argue that the human right to water is necessary for several reasons, most notably the non-substitutability of drinking water, rendering this right "essential for life." Some would contend that the right to water already implicitly exists; for example, it is recognized through legal precedents when courts support the right of nonpayment for water services on grounds of lack of affordability.[38] In addition, other human rights that are explicitly recognized in the UN conventions are predicated on an assumed availability of water (e.g., the rights to food, life, and health); the human right to water would enable the fulfillment of these related rights.[39]

Debating the Human Right to Water

Explicit recognition of the right to water would strengthen its basis in international law, supporters argue, with several advantages. First, new legal avenues would be created, enabling citizens to use legal means to compel states to supply basic water needs. Second, the human right to water offers stronger protection than the current essential services designation, at least when it comes to state provision; it requires, as stated in General Comment 15, that states create a national plan for the achievement of the right, which would include specific targets.[40] Third, the human right to water would constrain state activity with respect to privatization through creating a series of requirements with respect to state capacity (e.g., regulatory capacity), governance (e.g., waiving of corporate confidentiality), affordability (e.g., maximum thresholds for water bills relative to income), and accessibility (e.g., specifying a geographical dimension to connection targets, such that new connections are oriented toward slum areas) that in turn might constrain or redirect private sector activity.

Opponents have pointed out several potential difficulties in implementing a right to water. These concerns echo general criticisms of the rights-based approach to development, which (critics argue) is overly prescriptive and unhelpfully insistent on tackling all issues at once (rather than prioritizing, given limited resources and reasonable implementation times).[41] Pragmatic (and often, but not always, conservative) critics reject the rights-based approach to development on the basis that "efficiency" (often allied with the rhetoric of "basic needs" or "essential services") will meet people's needs better than rights-based approaches. Underlying this critique is the claim that advocating a human right to water is satisfying but trivial, because it "belabors the obvious, and ignores what is difficult."[42]

One of the central difficulties pertains to the definition, in practical terms, of a human right to water. The UN committee defined the right to water as follows: "The human right to water entitles everyone to sufficient, safe, acceptable, physically accessible and affordable water for personal and domestic uses." But how much water, in practice, does this imply? The World Health Organization cites a minimum of twenty liters per person per day for survival needs, fifty liters for personal and domestic needs such as bathing, and one hundred liters as a desired level.[43] But water is a multipurpose resource and is required for many other activities (such as food production and preparation, industrial production, and environmental protection) in widely varying amounts. Moreover, this definition does not enable us to adjudicate between competing uses (particularly in water-scarce regions). The UN committee's relatively narrow definition does not address these broader issues.

Another difficulty arises with respect to issues of accountability: there is no obvious way to hold governments accountable for implementation (particularly since oversight mechanisms for economic, social, and cultural rights are weaker than for corresponding mechanisms associated with civil and political rights).[44] Similarly, the human right to water raises the issue of justiciability: it is difficult (as with many economic, social, and cultural rights) to imagine translating a human right to water into an enforceable law; its interpretation (as suggested by the case study of South Africa, to follow) will be hotly contested. Moreover, a human right to water is likely to be regarded, in practice, as "soft law," easily trumped by the "hard-law" requirements of bilateral trade and investment treaties.[45] This is particularly the case given that investment arbitration tribunals (under which disputes over international investments, such as those pertaining to private water contracts, are adjudicated) have shown themselves to be reluctant to discuss the human-rights aims of states; this reticence implies that governments cannot be certain that actions that fulfill the state obligation to provide a human right to water will not be in breach of other obligations under investment treaties.[46]

The question of pricing is equally contentious: some plead that water should be free (Agenda 21), others that it should be affordable (the UN committee's position), and others that water should be priced at full (or sustainable) cost. The difficulties with working out these distributive issues, in practice, are well illustrated by the following discussion of South

Africa—one of the few countries to take implementation of the human right to water seriously. The South African case suggests that a right to water will result in little practical change: the right to water enshrined in South Africa's postapartheid constitution, for example, has not prevented large-scale disconnections and persistent inequities in water distribution.[47] In other words, a human right to water does not imply that water should be accessed free of charge (although it might imply an affordable basic "lifeline" supply within a reasonable distance from the home, or adapted to the capacity of the household to secure water supplies),[48] although this policy is at odds with cultural and religious views on water access in many parts of the world.[49]

Environmental issues are another source of concern. Critics warn of potential abuse of the concept of the human right to water, as governments could over-allocate water to privileged groups, at the expense of the environment. The human right to water also fails to consider in-stream uses and other water protection policies.[50] Human rights, in other words, are invariably anthropocentric. The implication arises that, in failing to recognize rights of nonhumans (or ecological rights), the provision of a human right to water may, ironically, imply the possibility of further degradation of hydrological systems on which ecosystems (and thereby human beings) depend.

Finally, the human right to water may conflict with existing systems of water rights, in both developed and developing countries. The potential legal tensions between human rights in developed countries and existing systems of water rights (for example, riparian rights and prior appropriation rights in North America) recognized by states is one issue. But many systems of water rights are not formally recognized by states; these are often territorially based, local in nature, and of crucial importance to indigenous peoples.[51] The negative implications of the creation of water markets in rural areas for indigenous water users in the Andes, for example, have been well documented.[52] Indeed, some indigenous-rights activists fear that a human right to water will provide additional leverage for states intent on wresting control of water resources from local communities. This is echoed by the arguments of scholars who critique the Western bias (or Eurocentrism) of human rights, which (from their perspective) stems from an individualistic philosophy that is culturally conditioned, rather than universal in nature. As such, they say, human rights are of limited applicability in non-Western societies. In the case of water, for example, complex cultural relationships with hydrological landscapes imply highly differentiated (and usually collective) sets of water rights only imperfectly captured, and indeed sometimes incompatible with the notion of water as a human right.[53] Others respond that the human right to water, like other rights, is an important mechanism that enables individuals to make claims against states, and a critical tool in the fight for social justice, regardless of cultural context.

From the perspective of antiprivatization campaigners, an additional criticism of the human right to water is that it will not foreclose private sector management of water supply systems. Current interpretations suggest that private sector provision is compatible with the human right to water.[54] The most important piece of international jurisprudence to date (the 2002 General Comment 15 by the Committee on Economic, Social, and Cultural Rights discussed earlier) acknowledges that states have a responsibility to ensure the human right to water but refrains from commenting on the legitimacy of different political-economic models of provision. This position is in alignment with human-rights law norms, and refutes the claim, made by some antiprivatization activists, that private provision violates the human right to water.[55] In other words, the human right to water does not foreclose, and indeed is compatible with, private ownership of water supply infrastructure (and, indeed, private rights to water resources).

The shortcomings of invocation of the human right to water by antiprivatization campaigners became apparent following the Kyoto World Water Forum in 2003, as proponents of private sector water-supply management began speaking out in favor of water as a human right. Senior water-industry representatives identified water as a human right on company websites, in the media, and at high-profile events such as the Davos World Economic Forum.[56] Right-wing think tanks such as the Cato Institute backed up these statements with reports arguing that "water socialism" had failed the poor, and that market forces, properly regulated, were the best means of fulfilling the human right to water.[57] Nongovernmental organizations closely allied with private companies, such as the World Water Council, also developed arguments in favor of water as a human right.[58] Shortly after the Kyoto meeting, the World Bank released a publication acknowledging the human right to water.[59] And less than a year after the Uruguayan parliament approved a constitutional amendment to the human right to water, the government passed an executive resolution allowing the private companies that had signed concession contracts prior to the referendum to continue.

Two years later, at the 4th World Water Forum in Mexico City in 2006, representatives of private water companies issued a statement recognizing the right to water, and recalling that the private sector had officially endorsed the right to water in 2005 at the thirteenth session of the UN

Commission on Sustainable Development.[60] At the Mexico City forum, a contrived consensus across civil society, the private sector, and governments on the right to water—albeit a diluted version—emerged.[61] Despite dissenting views of some governments (notably Bolivia), the diluted interpretation of the human right to water prevailed in the Ministerial Declaration of the forum, in regards to which private companies had an officially sanctioned role.

Ironically, this has occurred at the same time private companies have acknowledged the significant barriers to market expansion in the water supply sector in lower-income countries. As explored in chapter 3, analysis of the public statements of senior executives at water supply firms reveals a retreat from earlier commitments to pursuing private sector opportunities in water globally, with senior figures publicly acknowledging high risks and low profitability in supplying the poor.[62] Some international financial institutions have begun officially acknowledging the limitations of the private sector.[63] High-profile cancellations of some of the world's largest water-supply concession contracts—including Atlanta, Buenos Aires, Jakarta, La Paz, and Manila—seem to bear out the hypothesis that water presents significant challenges to private sector management. The multinational private sector has indeed retreated from supplying water to communities in lower-income countries, but as detailed in chapter 3, this has been due to the failure to achieve acceptable returns on investments and control risks, not solely (and in many instances, not primarily) to antiprivatization, pro–human rights campaigns. Companies continue to insist that water is a human right, one that they are both competent and willing to supply if risk-return ratios are acceptable—a condition that cannot be met by most communities.

This outcome is not unexpected by critics of human-rights doctrines operating from what we might term a radical left tradition, which posits that human rights are compatible with capitalist political-economic systems. They would point out that many citizens of capitalist democracies accept that commodities are not inconsistent with human rights (such as food and shelter), but that some sort of public, collective safety net must exist if these rights are to be met for *all* citizens. This requirement is true for housing and food (as inadequate as these measures may be in practice). Full privatization may be inconsistent with a human right to water unless it is coupled (as it is in England) with a universality requirement (laws prohibiting disconnections of residential consumers) and with strong regulations for price controls and quality standards.[64] But private sector *participation* in water supply certainly fits within these constraints. In other words, rooted in a liberal tradition that prioritizes private ownership and individual rights, the current international human-rights regime is flexible enough to be fully compatible both with private-property rights and with the private provision and operation of infrastructure, whether for water or other basic needs. This "tyranny of rights," as Brewster Kneen terms it, has important limitations in the defense of the public domain (whether conceived of as a "commons," "public goods," or the classic state provision of services).[65] In short, human rights are individualistic, anthropocentric, state-centric, and compatible with private sector provision of water supply; and as such, the human-rights argument is a limited strategy for those seeking to refute water privatization or the application of market principles to water management. Pursuing a human rights framework as an antiprivatization campaign thus makes three strategic errors: conflating human rights and property rights, failing to concretely connect human rights with different service-delivery models, and thereby failing to foreclose the possibility of increasing private-sector involvement in water supply.

CONTESTING A HUMAN RIGHT TO WATER IN SOUTH AFRICA

The contestation between consumers and the South African government over the human right to water and access to basic water supply illustrates the issues I have been discussing. Despite having enshrined a right to access to water in its 1994 postapartheid constitution,[66] postapartheid South Africa has been plagued by public protest over its water supply policies over the past decade. Access to water and sanitation services remains highly correlated to wealth and race. Under apartheid, differential access to water was institutionalized through legal mechanisms (which, for example, linked access to water to land rights) and governmental structures, and materially entrenched by differential access to domestic water-supply networks. Water resources were allocated and water law was "written largely in the interests of commercial agriculture and industry...and [evolved] largely to meet the needs of changing industrial, metropolitan, and agricultural users which necessarily excluded a large sector of the population."[67] Whereas white communities continue to enjoy "first world" services, the majority of black South Africans have access to services of much poorer quality, or no access to water and sanitation whatsoever; in 2001, 3 million households were without access to the legal minimum of twenty-five liters of potable water per person per day within two hundred meters of their home.[68]

The delivery of basic services such as water and electricity to communities denied those services under apartheid has been one of the core platforms of the African National Congress government since the end of apartheid rule in 1994. Improving access to water, however, has been complicated by the high operating and capital costs, and by ongoing changes in governance within South Africa—notably the complex process of capacitating (and sometimes creating) local government structures.[69] Although responsibility for water supply retail services was shifted to local governments in the late 1990s,[70] critics have charged that this represents the transfer of an "unfunded mandate" for water supply and sewerage to local government. The situation a decade after apartheid remains one of stark differences in access and use. Domestic water use rates by wealthier households can average six hundred liters per person per day, whereas rural communities and township residents without household connections may average ten liters per day.[71]

Faced with these high costs, and the large extent of unmet demand, the national Department of Water Affairs (originally charged with addressing the "backlog" in water and sanitation services) began implementing a goal of clean, affordable water supply of twenty to thirty liters of water per person per day, at a distance of no more than two hundred meters from the home. Government policy on payment for these services was clear from the outset: a human right to access water did not imply access to *free* water. This is bolstered by reference to the constitutional court's ruling that the steps taken to provide water are "reasonable," with specific reference to the roles and capacities of governments as well as the requisite resources.

Access to water in South Africa is therefore enhanced by constitutional rights but is by no means guaranteed, because economic and political factors shape what is considered "reasonable." In the mid-1990s (shortly after the end of apartheid), for example, individual household connections were metered using prepayment meters (banned in countries such as the United Kingdom), and punitive cost-recovery measures were implemented for defaulters, frequently cutting off electricity and water supplies simultaneously.[72] An example cited by the Department of Water Affairs and Forestry (DWAF) in a manual detailing "Successful Cost Recovery Case Studies for Water Services" makes the severity of cost-recovery measures clear:

Defaulters were dealt with harshly. Accompanied by police in armored vehicles, municipal workers were sent into settlements and put in flow restrictors so that defaulters, with their taps open 24 hours per day, got only about 15 Kilolitres per month. At the same time the electricity is cut off. Initially there was intimidation, but community meetings were held and the situation discussed. Once their debt has been paid, defaulters have to pay R20 to have the flow restrictors removed. Persistent defaulters are handed over to lawyers for legal action. After about 3 months people had come to accept the inevitability of payment.[73]

Yet this inevitability of payment is by no means easily accepted by people in many South African communities who feel that they have gained nothing from the new political dispensation, and yet are expected to pay for services previously received as public goods under apartheid. In both urban and rural areas, there exists a well-established tradition of resisting authorities, boycotting service charges, and opposing the enforcement of disconnections to utility services. This resistance contributed to massive local-government debts in the final days of apartheid, forcing the national government to subsidize local governments in order to meet rising deficits in urban areas.

Accordingly, South African governments since the end of apartheid have been hostile to nonpayment and have initiated numerous campaigns to counter the practice, including disconnections. Despite this, arrears for water bills have risen rather than fallen over the past decade.[74] Hostility toward ANC policies was felt at the local level in the townships, and additional political pressure came from white neighborhoods as rates have risen to cover some of the costs of providing services in neighboring townships, with the Sandton ratepayers' boycott perhaps the most notorious. One group of Pretoria residents refused to pay water and electricity bills to the city council after amalgamation with neighboring townships, claiming that the higher rates charged by the council to (predominantly white) residents of "Old Pretoria," effectively cross-subsidizing services to the townships, constituted a violation of their constitutional right to equality. Pursuing their claim all the way to South Africa's Constitutional Court, the residents lost their case,[75] but protests such as these underscored both the political dangers of cross-subsidization and the necessity of forcing cost recovery on low-income South Africans with household connections, despite their poverty and public-health considerations, such as the estimated 11 percent HIV-prevalence rate.[76]

On the basis of these policies, disconnections of water customers for unpaid debts continued in some areas throughout the latter half of the 1990s. Campaigns to encourage payment for water services also continued. The cholera outbreak in KwaZulu-Natal in 2000–2001 demonstrated the dangers of this approach.[77] In some communities, this was the first time people had been asked to pay for water supply; rather than pay, they turned to traditional water sources, such as rivers, streams, and ponds. In some areas, water supply was cut off to residents; they, too, turned to traditional

sources, where they contracted the disease.[78] As a result, the cholera epidemic quickly spread, killing more than two hundred and infecting an estimated one hundred thousand.[79]

Belatedly, and under intense criticism, the national government announced a new "free basic water" policy, a sharp departure from the full
cost-recovery policy it had maintained throughout the latter half of the
1990s with the active encouragement of international financial institutions.[80] Yet the six thousand liters per household per month free of charge
(which translates into twenty-five liters per person per day, for an average
black South African household of eight) was associated with new, steeply
rising block tariffs that resulted in high water bills for those consuming over
the free limit. Moreover, "free basic water" was only to be made available
to households not in debt or arrears on water payments, while the practice
of water cutoffs for nonpayment of electricity bills continued in some municipalities.

Simultaneously, some municipal governments in South Africa began new
metering, pricing, and conservation programs, designed to improve cost
recovery while encouraging reductions in water use (and water wastage).[81]
The program undertaken by the City of Johannesburg in Soweto (one of
Johannesburg's largest, poorest, and mostly black townships) was typical:
it sought to reduce water leakage and to improve payment for water services, given that only about 10 percent of households paid their water bills.
The city government felt that it had good reasons for implementing this
program. Like many other water utilities in postapartheid South Africa,
Johannesburg Water[82] faced serious revenue shortfalls; it also had worries
about future supply, given rising consumption and expectations in this arid
country. It seemed reasonable that Soweto—with poorly maintained infrastructure (and consequent low pressure and high leakage rates) and low
levels of bill payment—would be the city's target. Residents were offered
two choices: unrestricted water supply, subject to accepting a prepayment
meter, which supplied a basic amount of water free, and beyond which
people had to buy tokens to pay for consumption in advance; or restricted
water supply at no cost, but with indoor household taps removed, and an
outside standpipe fitted with a flow-restriction device to limit the amount
of water they could receive. For many residents, this meant that for the
first time they had no option but to pay for the water they used, or to get a
downgraded and inconvenient outdoor connection.

But the hardship on poor families resulting from Operation Gcin'amanzi
(Operation Save Water), as the city called it, resulted in ongoing community protest. In 2006, a court case was initiated by five female residents

of the Phiri neighborhood of Soweto, supported by the Coalition against Water Privatization.[83] All five applicants were unemployed and lived in large households—in some cases including relatives living with HIV/AIDS who required extra water for basic medical care. Previously supplied with unlimited water for which a flat fee was levied, all five residents, together with thousands of their neighbors, either had their water supply cut off by Johannesburg Water as part of Operation Gcin'amanzi or were persuaded to accept a prepayment meter. Without water for as long as two weeks per month, the applicants asserted that twenty-five liters per day was insufficient for basic household needs. Supported by expert testimony, they demanded that the court declare unlawful the city government's limited water access and punitive cost-recovery policies, asking for fifty liters per person per day and a credit-metered supply option. In April 2008, the high court ruled in favor of the Phiri applicants: prepayment water systems were unconstitutional and unlawful, and the city should provide residents with fifty liters of water per person per day. The city immediately appealed the decision.

As the testimony of government officials and managers of Johannesburg Water indicates, however, the municipality and water utility have little scope to meet the demands of Phiri residents. On the one hand, the city invoked fears of water scarcity, arguing that rapid increases in water use by poor residents would exceed the city's available water supply.[84] Prepayment meters would promote much-needed water conservation, argued the city in a press release following the high court's decision. But the call by activists for new tariff structures to apply to the high-volume consumption by wealthier Johannesburg residents—who use hundreds of liters of water per person per day—remained largely unacknowledged by the city.

The City of Johannesburg also pointed to economic constraints under which it was operating. Stating that the water utility's approach was "reasonable" and "respected the rights of residents," and noting the large financial burden placed on municipalities by the legal obligation to extend water connections to all households, defendants from Johannesburg Water argued in line with South African legislation that the requirement of an "economically sustainable" water supply meant that larger quantities of free water could not realistically be supplied, and that meeting the minimum requirements specified in central government policy was sufficient for meeting the constitutional right of residents.[85] As noted by the water utility's legal counsel, a human right to *access* water does not imply a human right to *free* water.[86]

Human rights, from this perspective, place only a limited onus on governments, do not preclude private provision, and are not inconsistent with the application by the public sector of market principles such as full cost recovery for the provision of services. The implication follows that if residents cannot pay, they cannot avail themselves of their human rights. As Phiri residents pointed out, however, cross-subsidies could provide a mechanism for providing more equitable water services for all; but the politically contentious nature of this redistributive agenda has meant that the progressive tariffs called for by activists have not been implemented.

These debates are far from settled in South Africa. The supreme court of appeal overturned the high court's ruling in 2009; on appeal, the constitutional court upheld the ruling in favor of Johannesburg Water. And more restrictive conditions were imposed: although the high court ordered the city to provide residents of Phiri with fifty liters of free water per person per day and to remove prepayment meters, the supreme court of appeal ruled that forty-two liters per person per day was sufficient and ruled in favor of meters.[87]

Although their appeal was rejected by the constitutional court, activists continue to argue that prepayment meters are illegal and that a "reasonable" amount of water, particularly for those in large households and those living with HIV/AIDS (of epidemic proportions in South Africa), is well beyond the fifty liters per person per day adjudicated by the high court. The result is a stalemate: while activists claim "free water" as a legal right and call for the removal of water meters by residents themselves (an action which the city characterizes as illegal), the city continues to defend water meters and water tariffs as a legal (and indeed the only practical) solution for supplying "water to all" in South Africa. The city continues to attempt to develop a "social package" to provide more relief to poor and vulnerable households and to design strategies to balance the redistributive effects of its tariffs and policies between wealthier and poorer households. And activists continue to argue that the city's redistributive agenda does not go far enough. The existence of a human right to water has not, in South Africa, led in any straightforward sense to "water for all," but has rather opened up a debate on how to supply water affordably and sustainably over the long term, on the basis of a widely accepted definition of equity.

CONCLUSIONS: BEYOND THE HUMAN RIGHT TO WATER

The South African example illustrates the advantages and disadvantages of a human right to water. Difficult to define and implement, with potentially ambivalent consequences for environmental sustainability, the human right to water is no panacea: it leaves many questions unanswered. Can a human right to water be practically realized? If so, what institutions and legal mechanisms might be required? What are the environmental and social consequences of an unlimited right to water? How can ordinary citizens hold to account governments that fail to adhere to laws aimed at ensuring water access and sustainability? The South African case suggests that these questions do not have easy answers.

The case also suggests a range of criticisms that might be made of the human right to water. Foremost, the framework of human rights is individualistic and legalistic, and hence cannot address the complex, collective governance issues that constrain access to water at the urban scale. Equally, it cannot foreclose private provision of water supply or the commercialization of water, although it does provide an alternative perspective from which to analyze the implications of commercialization. The equitable provision of water supply necessarily implies a degree of solidarity, both physical and material. Yet it is precisely this notion of solidarity that human rights, in isolation, cannot provide. Equally, the anthropocentrism of human rights fails to recognize rights of nonhumans (or ecological rights). Providing a human right to water may, ironically, imply the further degradation of hydrological systems on which ecosystems—and human beings—depend. And while a human right to water is consistent with public and private models of water supply provision, it fails to incorporate (and indeed may lead to violations of) the norms and commitments of community-controlled water supply systems. In other words, "rights talk" resuscitates a public-private binary that recognizes only two options—state control or market control—that, for reasons explored in part 1, are in many ways equally unsatisfactory.[88]

Moreover, "rights talk" offers us an unimaginative language for thinking about issues of water privatization, not least because it reinforces the public-private binary on which this confrontation is predicated, occluding possibilities for collective action apart from corporatist models of service provision. In other words, one might reasonably conclude that the discourse of human rights serves, in a narrow interpretation, not to resist but to legitimate what opponents term "neoliberalism": the doctrine that the market exchange is an ethic that should serve as a guide for all human action (see chapter 1). This critique suggests that human rights provide legal sanction to moral claims, but only in a limited fashion, and in such a way as to render the concrete realization of these claims practically impossible. This dilemma emerges because legal rights are indeterminate, and hence are inevitably transformed into bureaucratic, technical formulations (such as a "minimum" amount of water per person per day that satisfies the human right to drinking water) that evade broader questions of social justice.

The critiques have some validity, as the ongoing debate in South Africa suggests. Nonetheless, I would argue that the human right to water is a useful tactic for those without access to legitimize their struggles, not only for water, but also for human dignity. In this aspirational sense, the human right to water is a valuable tool. A human right to water compels the powerful to consider the redistributive politics embedded in their water use practices. In other words, human rights are not the solution but are rather a strategy for creating the context in which claims for social justice can be pursued. As Malcolm Langford argues, "The history of human rights shows that once people feel and experience something as a human right, it becomes a difficult force to restrain."[89] Alternative approaches (such as "basic needs" or "essential services") favored by proponents of commercialization do not lend as much legal weight to social-justice considerations. Indeed, as explored in earlier chapters, the mechanisms of public participation and decentralization often associated with commercial approaches can, in some instances, render the practical achievement of equitable water access more difficult.[90]

In this sense, human rights are a useful antidote to what some scholars term "elite capture" of the benefits of development.[91] Mobilizing around human rights can foster solidarity and the construction of collective identities. This is particularly the case, as Bronwen Morgan argues, when legal strategies (such as the pursuit of a human right to water) are combined with illegal strategies (such as disconnecting water meters, and unauthorized connections to water supply networks).[92] In other words, although a narrowly legalistic approach to "rights talk" is insufficient, it can have political potentiality if interpreted more broadly.

The human rights–based approach to development, in other words, may be helpful because it confers "moral legitimacy and the principle of social justice," particularly in the realm of subsistence (or, as some prefer to term it, "social consumption").[93] A rights-based approach asserts that inequalities—economic and political—play a central rather than marginal role in the production and reduction of poverty. This compels us to examine issues of political and social power, and the processes of negotiation and contestation (formal and informal) by which the costs and benefits of development are shared; the need for participation, accountability, legal enforcement, and a focus on marginalized groups all flow from this perspective. But, as Morgan notes, maintaining this progressive potential is difficult; the case of South Africa illustrates the difficulties of institutionalizing the gains from activism around the human right to water. Only through maintaining a broader sense of the moral claims underpinning human rights (such as Mitchell's call for the "right to the city," or critical legal scholars' invocation of the "right to have rights") can "rights talk" be mobilized effectively in support of social justice claims.[94] Effectively realizing human rights, by implication, requires the articulation of property rights and water rights; in urban areas, as explored in chapter 4, this implies a link between water access and land tenure, and between water use and land-use practices.

This approach obviously has much in common with the expanded notion of governance explored in previous chapters. To briefly summarize: while government agencies often recognize that unserved communities have legitimate claims to social services, they are frequently unable to provide those services, and so deal with their claims on the basis of political expediency. In order to collectively apply informal pressure on governmental actors to meet their demands, these marginalized populations seek to define themselves as communities with a clear identity (such as "slum dwellers" or "water sellers").

The recognition of these groups as communities, as scholars like Appadurai and Chatterjee argue, is critical to enable claims to livelihood and public services to be made. In other words, the demands of these communities are often put forward in the name of a collective (sometimes long-standing, but sometimes recent in origin, as with many informal urban settlements). The social or consumption rights these groups demand are thus distinct from human rights in general, which focus on individual rights characteristic of Western liberal-rights frameworks. This struggle is sometimes termed the "right to have rights": a struggle to expand the notion of citizenship to include marginalized and excluded people.[95] (Recent academic debates over social citizenship make a similar point.)

To the extent that human rights facilitate this latter type of struggle, they are congruent with the radical versions of rights called for by some critical legal studies scholars. These radical formulations of human rights subvert the norms of what Charles Taylor terms our (Western) "modern social imaginary": neatly separated spheres of public and private, the economy as a socially and environmentally dis-embedded system of exchange, the stability of property as a social relation, and the universality of civil society (with its presumption of the equal capability of all citizens to engage in processes of popular sovereignty, under the aegis of a benign state).

The human right to water is currently advanced as the best strategy of defending community access. In this sense, it is no accident that antiprivatization activists call themselves "water warriors": they are invoking a long-standing discursive opposition between the ethos of commerce, on the one hand, and the ethos of civic responsibility, on the other.[96] But their defense (although rarely coherently articulated) is really of a collective of water users (however ill defined) and the environmental conditions that sustain them, rather than individuals as conceived in conventional human-rights frameworks.

It is thus unsurprising that opponents to privatization often assert that their concern is with the tension between corporate and community control, and not merely the choice between public and private management. Demands for the extension of water supply to poor households are attempts to deepen democracy—and control over policymaking by marginalized communities—by making claims over material emblems of citizenship: the basic public services necessary, particularly in urban areas, for dignified human lives. A logical consequence of the expanded struggle for the "right to have rights," in other words, is the claim that community forms of water governance have the potential to be inclusive and environmentally sustainable—a viable alternative, that is, to both government and private provision. This view underlies the claims of activists and scholars who invoke alternative concepts of rights: property rights (most frequently some form of commons) and morally informed water rights (in which the nature of water precludes private ownership, usually on spiritual or environmental grounds, and requires forms of trusteeship or stewardship administered by collective forms of social organization, whether communities or the state). This in turn implies the need to examine other management options, including alternative business models, governance models, financing, and technologies. The next two chapters turn to an exploration of these alternatives, with reference to more general debates about the viability of alternative governance models for supplying water.

6

COMMONS VERSUS COMMODITIES

THE AMBIGUOUS MERITS OF COMMUNITY

WATER-SUPPLY MANAGEMENT

The past few years have witnessed a surge of calls in support of community water supply. Appeals to the "water commons," calls for "water democracy," and campaigns for the community control of water supply are examples of the central role that community (however ill defined the term might be) plays in the vision of antiprivatization activists.[1] The concept of community has also come to play a greater role in mainstream development policy, even on the part of proponents of the commercialization and privatization of water. Although community provision was largely marginal to the early arguments of those in favor of water privatization, a resurgence of notions of community has occurred over the past decade in discussions of urban water supply. These ideas of community have taken various—and not always compatible—forms, from public-private-community partnerships, to community-based private-sector water providers (such as water vendors), to community business partnerships.[2] These disparate ideas of community are often mobilized together with participatory governance strategies (variously labeled "delegation," "comanagement," or the like) and commercial water-management approaches. The term "community" (and its cognate, the "commons") is thus frequently mobilized by those on both the political right and left—a seemingly paradoxical convergence that engenders an ambiguous politics and makes the evolution of commons and community water-management regimes difficult to predict and interpret in practice.

This renewed interest in community echoes the broader resurgence of interest in the so-called environmental commons, which has become an important strand of thinking within environment and development debates.[3] Rather than defending government or market provision, advocates of the commons argue that they seek an alternative to both states *and* markets, as captured in the opening lines of a chapter on the commons in the 2008 edition of the Worldwatch Institute's influential *State of the World* report:

> For decades we have been told that there are only two choices for the management of scarce resources: corporate self-seeking or the

bureaucracy of the state. But there is another way. Commons management has worked for centuries and is still working today. It can be adapted to the most pressing global problems, such as climate change. A new phrase is about to enter the policy realm. To "market-based" and "command-and-control" we can now add "commons-based."[4]

This invocation of the commons shifts the terms of debate. Rather than a confrontation between states and markets, we oppose the commons, on the one hand, and corporate forms of ownership and management (whether state or market), on the other. These are, of course, excellent political tactics, particularly in the United States, with its tradition of antigovernment sentiment, but also internationally, insofar as this framing of the commons has the potential to unite diverse struggles, ranging from rural farmers in developing countries calling for land reform to Internet hackers arguing for a global cyberspace commons and environmentalists arguing for alternative approaches to atmospheric emissions control. In this sense, the popular debate on the commons has diverged from mainstream academic debates, particularly those inspired by the work of Elinor Ostrom, which have tended to focus on commons management as distinct institutional forms that are not in direct or necessary opposition to states or markets. (My formulation of community versus corporate (state/market) provision has, perhaps, more in common with the work of Robert Wade.[5])

Appeals to the commons of this sort (and the implicit call for community management with which they are often associated) speak to various conceptual debates—over the post–Washington Consensus, postneoliberalism, or Third Way approaches—that seek to rearticulate the relationship between governments, markets, and communities. In the wake of the private sector retreat from the water supply business in much of the world (and the global financial crisis that had hit with full force by the end of 2008), calls for alternative community management proposals have gained momentum. But, as we shall see, they are at times deeply flawed, and less widely applicable than their proponents might admit.

So what are we to make of calls for commons and community water management?[6] Should we pursue or dismiss them? And what might they mean in practice? In this chapter, I propose to answer these questions, with reference to the broader debates over governance discussed in earlier chapters. At the outset, it is important to note a distinction between approaches to the commons in public debate. On the one hand, we have institutionalist (which tend toward managerial and technocratic) definitions of the commons, more often invoked by the Right, which are associated with quite specific organizational and operational characteristics (often not that different, in practice, from market-based management regimes). On the other hand, we have communitarian interpretations of commons and communities, usually invoked by the Left, which are often associated with specific political and ethical stances (e.g., anticapitalist, appropriate technology) but more rarely address detailed institutional issues. We might term the former a set of "blueprint utopias" and the latter a set of "fictional utopias." The uneasy dialectic between these two broad concepts of the commons, and the ambivalence of their implementation in on-the-ground struggles over water management alternatives, are central themes of the chapter.

I should note, however, that my opposition between these two approaches—institutionalist versus communitarian, fictional versus blueprint—is not meant to imply a straightforward correspondence with pro- or antiprivatization views. Those of either institutionalist or (although more rarely) communitarian persuasions can be pro- or antiprivatization, although for very different reasons. Rather, this distinction will serve to highlight tensions in the debate over commons and community water governance in the context of water privatization. This debate, as we shall see, differs from the mainstream literature in its definition of the commons: the distinction between corporate (state or market) and community forms of resource management is more pronounced; and it is often associated with a distinction between commons and commodities that is absent from the mainstream academic literature on the commons (and, indeed, would be viewed from a mainstream perspective as analytically incorrect).

To draw out these themes, I begin with a short, illustrative parable of Cochabamba, Bolivia—home to (perhaps) the most well-known antiprivatization movement over the past twenty years. My use of the term "parable" is deliberate: this is not a case study, but rather a simple tale designed to introduce and contextualize the key questions with which I wish to engage. The second part of the chapter presents a typology of community alternatives, drawing on (although not in a comprehensive fashion) the extensive literature, both academic and policy, that exists on community water management around the world. This is followed by critiques of various conceptualizations of community, with reference to more general debates about the viability of the commons (or, more correctly, "common-pool management regimes") and communities to manage and govern public goods such as water. Throughout, I attempt to approach water and its multiple uses in a relational fashion, focusing on water supply as articulated with other forms of water services, such as irrigation.

Before going further, I should insert a caveat. This chapter does not engage extensively with specific institutional forms that the commons and community water governance take in peri-urban areas. This is, in part, because a detailed description of these forms was presented in earlier chapters (particularly chapters 1 and 4). In those earlier chapters, I examined the interactions between these multiple forms of community provision, states, and markets. The present chapter builds on this earlier analysis to focus on the conceptual underpinnings of the debate over community water supply. As a conceptual analysis, it does not engage with the specifics of particular community approaches, and leaves aside discussion of technical issues (such as regulation) and political issues (the relations of power and clientalism that structure community water supply systems around the world). My entry point into this discussion is not technical or political: rather, I begin with the disjuncture between discourse and practice around community water supply, and then expand on the simple observation that communities often—indeed perhaps always—operate in ways that undermine the straightforward storylines of community management proponents.

RETHINKING COMMUNITY: COCHABAMBA'S "WATER WAR"

Cochabamba's *guerra del agua,* or "water war," has become emblematic of the potential power of social movements, and as such is celebrated by antiprivatization activists as a victory of community over private capital. Yet a closer examination of Cochabamba's experience following the water war suggests that there exist significant limits on the power of communities to improve water-supply access for the urban poor. It also provides an example of the possibilities and potential pitfalls of community-led water governance—which is currently widely supported by both proponents and opponents of water commercialization.

The main events of the water war are well known.[7] Privatization inspired mass protests that became a cause célèbre for alter-globalization activists. Access to water had long been a contentious issue in Cochabamba, Bolivia's third-largest city. Wealthy consumers and businesses received municipally subsidized water through the network, whereas half of the city's households—largely indigenous residents of poor neighborhoods—relied on water delivered by tanker trucks, private wells, or water systems that were small scale and often community run.[8]

The causes of Cochabamba's lack of water supply were complex, observers admitted: a lack of new water sources, overconsumption of water by wealthy households, the disinterest of the city mayor and political elites in extending the water supply system to the poor, inefficient management, and a culture of political exclusion of indigenous farmers (*campesinos*) all played a role.[9] Debates over potential solutions were foreclosed by the advent of "shock therapy" in the mid-1980s, as Bolivia began a series of IMF-administered structural-adjustment programs.

In 1998, the IMF made a new loan to Bolivia, contingent on privatization of a range of state enterprises, including Cochabamba's water supply system. That same year, the Bolivian government passed new laws that enabled incentives to be offered to private water companies to manage public water supply systems, and granted those companies rights to rural water supply sources that had traditionally been under the control of indigenous farmers. An additional source of pressure came the following year from the World Bank, which recommended the elimination of subsidies to Cochabamba's water supply system. The reward, promised proponents, would be an extension of the network to those without access. Events were to prove them wrong.

The private consortium that signed the forty-year concession contract with the city government in 1999 included a subsidiary (based in the Cayman Islands, an international tax haven) of two of the largest water companies in the world: the U.S.-based multinational Bechtel and Britain's United Water. As is frequently the case in water privatization deals, this local subsidiary, Aguas del Tunari, was able to negotiate a highly favorable contract (in this case, aided by the fact that it was the sole bidder). One clause of the contract guaranteed a profit of 15 percent to the consortium; another indexed the profit rate to foreign currency exchange rates, as a protection against devaluation of the Bolivian currency. The contract also gave the company exclusive rights to all of the water in the Cochabamba Valley, including rainwater, and private wells in rural areas surrounding the city. This was an unprecedented move, and unusual given that in the intensively farmed Andean mountain valley in which Cochabamba is located, wells and streams are essential sources of drinking water and irrigation for indigenous *campesinos,* who managed local water sources largely independent of external control.

Shortly after taking control of the water supply system, the Bechtel subsidiary sharply increased water prices. Government sources and anti-privatization campaigners publicized cases of increases as high as 200 percent (although company officials maintained that average price rises were 35 percent). The company also undertook to place water meters on private wells and the local irrigation and cooperative water supply systems that rural and peri-urban residents had independently built and financed. Company managers likely underestimated the reaction of FEDECOR (the

Cochabamba Departmental Federation of Irrigators' Organizations), a powerful regional organization of irrigators formed in the early 1990s in response to municipal government attempts to drill deep wells to tap into new sources of water.[10] Given its links to urban social movements and its large base in rural and peri-urban areas, FEDECOR was to play a key role in resisting Aguas del Tunari, together with unions, women's groups, environmentalists, and consumers' groups. A key unifying factor for the diverse group of protestors was the sacredness of water and its cultural resonance for the indigenous population of the Andean highlands, allied with nationwide political struggles against the "neoliberal turn" underway in Bolivia.

Within months of the start of operations, it quickly became clear that the company had overstepped its bounds. A protest organization quickly emerged: the Coordinadora de Defensa del Agua y de la Vida (Coalition for the Defense of Water and Life). The coalition organized a city referendum in March 2000, through which the vast majority of residents expressed their disapproval of the privatization contract.

Widespread street protests began in April 2000. Tens of thousands of people took to the streets: protestors succeeded in mobilizing the city's residents and shutting down the city with general strikes and road blockades. The government's response was harsh: a military state of emergency was declared, several leaders of the coalition were arrested and jailed after midnight raids, and violent confrontations between the military and protestors left a seventeen-year-old boy dead and hundreds wounded. As the coalition continued to report events in Cochabamba to the international press (with coverage from *The Economist* and CNN), union leader Oscar Olivera was spirited out of the country to Washington, DC, where he addressed the April 16 anti–IMF/World Bank rally.

Back in Cochabamba, the protests persisted, and Bechtel's foreign managers (mostly U.S. and British) fled the city. Shortly thereafter, the contract was unilaterally rescinded in response to campaigners' demands.[11] Martial law was suspended; government officials resigned or decamped; and for a period the city was governed by its residents, through community assemblies presided over by the Coordinadora in a manner likened by sympathizers to the heady days of the Paris Commune of 1871.

La guerra del agua has been internationally touted as a David-versus-Goliath success, and as one of the best examples of successful grassroots resistance to private sector participation in water supply in countries of the South. The water supply company was taken back under government control. A subsequent campaign led to the reform of national water legislation, giving greater protection to rural irrigators, whose *usos y costumbres* (customary rights to and uses of water) now have greater legal protection. And an international campaign played a significant role in persuading Bechtel to withdraw its suit for US$25 million in damages from the ICSID tribunal (a World Bank Group affiliate) in 2006. In Cochabamba, meanwhile, there were growing demands of the antiprivatization protestors for community control of the water supply system, a movement that had wide appeal in this Andean society where rural water rights have historically entailed the right to democratic control over water resources.[12]

Yet the outcome of the *guerra del agua* has not been as positive as one might expect for Cochabambinos. Attempts by the Coordinadora to gain greater control of the city's water supply utility (SEMAPA) have been largely thwarted.[13] Initial demands for "social control" of the water utility following Bechtel's departure were eventually diluted into a proposal for social representation on the board of directors, previously staffed exclusively by professionals and politicians. But this did not attract widespread community support: two rounds of elections of community representatives to the board attracted fewer than two thousand voters in a city of six hundred fifty thousand. Traditional suspicion of government as a means for advancing elite interests has outlasted, it seems, the spirit of radical democracy inspired by the water war.

This has meant, in turn, a failure to achieve the Coordinadora's goals of improved transparency and greater responsiveness to community demands for universal access. Connection rates were less than 50 percent in 2006, six years after the water war. Corruption, inefficiency, and leakage continued to drain both financial and water resources from the utility. Observers disagreed about the causes. Was it (a) the continued control of the mayor over the water utility's budget, (b) the stringent conditions attached to the new loan issued by the Inter-American Development Bank, (c) a lack of sufficient finance, or (d) the absence of new water sources? Regardless, all agreed that SEMAPA remained unresponsive to the needs of the poor, in a pattern unchanged from the pre–water war era.[14]

The failure to radically change SEMAPA's behavior in the period following the *guerra del agua* was a combination of all of these factors. I would also add two more: the difficulty of constructing a viable model of communal ownership and social control, as originally envisioned by the water war's leadership; and the difficulty of enacting effective models of collective solidarity beyond the local scale (such as the public-public partnerships advocated by antiprivatization activists). The demands for communal property had grown, in large part, out of the experiences of two key groups of participants: small, rural irrigation associations and rural communities (with an interest in communal ownership); and urban water users with a desire for more cooperative, democratic forms of urban supply governance. In the former case, irrigation systems were often run entirely independently of government oversight and, particularly in the case of farmers' irrigation cooperatives, had deep historical roots. In the latter case, communities in the poor neighborhoods in the southern part of Cochabamba built their own systems, largely supplied by wells, and operated them via informal committees or councils elected by residents. In the aftermath of the failed reforms following the water war, these community-run water systems continued to operate. Some secured financing from external sources (the European Union, the World Bank) to build or extend their independent water-supply systems that still continue to be independently managed by users (even where water is purchased in bulk from the municipal water company). These systems are largely unregulated by the state: tariff setting, quality oversight, and other regulatory functions applied to formal network providers do not extend their reach this far.

How are we to interpret this outcome? On the one hand, the collective ownership of infrastructure and devolution of responsibility for water supply to self-organized users is consistent with local customs of community self-help. This is central to the social fabric of Andean communities and is often justifiably a source of community pride. On the other hand, the creation of parallel networks entrenches the fragmentation of the water supply system, creating two tiers of service with vastly unequal levels of state support. Wealthy areas of Cochabamba receive government-subsidized services of high quality, whereas residents of poorer areas of the city must directly engage with donors and mobilize volunteer labor to create more expensive systems for which operating costs must also be borne by these poorer communities. The Cochabamba case thus presents us with a dilemma: in celebrating community resourcefulness and promoting community-run water supply systems, we risk condoning both government inaction and corporate misconduct, while depriving low-income communities of the services and resources that wealthier communities already enjoy.

DEBATING COMMONS AND COMMUNITY WATER SUPPLY

The story of Cochabamba presented above is intended as a parable, a means of presenting a few simple lessons for consideration. Specifically, my story illustrates the disjuncture between discourses and practices of community water management, and the difficulties of enacting community control of urban water supply systems. It suggests that there are different types of commons—including irrigators' cooperatives in peri-urban areas, and community self-built water supply systems in poor neighborhoods—and thereby calls attention to the distinctions, made at the outset of the chapter, between water resources and supply, and rural and urban uses. It also emphasizes the consequences of the fact that "public" services in many countries are limited to the elite.

In the case of urban water supply, this fact of exclusion does not arise, I should emphasize, from community provision per se (although community provision may be characterized by exclusion). As Chatterjee has argued (and as the term "elite capture" discussed in earlier chapters suggests), exclusion is integral to the process of modernization undertaken by developmental states.[15] These states often lack the resources required to universally provide public services (like electricity or water) on which their legitimacy is predicated. This is complicated by issues of identity; social ruptures within postcolonial states have frequently meant that only a subset of the population were identified as full citizens, with the full set of rights and entitlements that one might expect. Moreover, strategies and technologies of development, formulated through interchange with the West and encouraged by development agencies, are not always appropriate to their new contexts.

Invoking community alternatives to privatization thus requires an examination of our (implicit) definitions of community, as well as the type of water supply in question. In doing so, it is important to distinguish between water supply systems and (raw) water resources, on the one hand, and between ownership and governance, on the other. Proposals for community alternatives to water supply are rarely sensitive to these differences. Table 6.1 presents a simplified typology of these community alternatives, distinguishing between community ownership and governance, and between water supply and water resources.

Critiquing the Commons

The best-known example of communal or collective water ownership is, of course, the commons, or more precisely, common-pool water resource management. Simply defined, common-pool resources are those from which it is difficult to exclude individuals (either through physical barriers or laws) and for which use by one individual can reduce the benefits available to others. Consequently, private-property rights are difficult or impossible to establish. Local collective-management institutions thus evolve

TABLE 6.1. *Water, the Commons, and Community Economies: A Tentative Typology*

	Example
"The Commons" (community ownership)	Water Supply Infrastructure • Water cooperatives (e.g., water cooperatives in Finland; Katko 2000) • Low-cost, community-owned infrastructure (e.g., Orangi Pilot Project, Pakistan; Zaidi 2001) Water Resources • Communal water rights (Narain, 2006) • Collective provision of irrigation (Shiva 2002; Wade 1994)
"Water democracy" (community-led governance)	Water Supply Infrastructure • Customer service boards and "customer councils" (Franceys 2006; Page and Bakker 2005) • Participatory budgeting (e.g., Porto Alegre; Baietti, Kingdom, and van Ginneken 2006) Water Resources • Community watershed boards (Canada; Alberta Environment 2003) • Sharing of irrigation water based on customary law "*usos y costumbres*" in the Andes (Trawick 2003; Boelens 2008)

Source: Adapted from Bakker 2008.

within communities to deal with common-pool resources. These are conventionally known as "common property (or pool) management regimes," characterized by exclusive rights-holding membership.[16] Common property water-management regimes are widespread; they exist in both developed and developing countries.[17] To give a simple example, *acequia* irrigation systems in Spain are community controlled: community members have carefully detailed rights and responsibilities involving all aspects of diversion, allocation, use, and renewal (recharge) of water resources.[18]

Significantly, common-pool resources like water are in this sense distinct from true public goods, which are nonrivalrous.[19] Consequently, common-pool resources are potentially prey to overuse and, perhaps, eventual destruction. When unregulated—that is, when access is completely open, as with the oceans or global atmosphere—this is exactly what tends to happen, as in Garret Hardin's "tragedy of the commons."[20] But most common-pool resources are not open to access by all. As Hardin later admitted, the "tragedy of the commons" is a misnomer. Common-pool resources are usually governed by rules and social relations that control access with the goal of limiting overuse.[21] Where it occurs, Hardin's tragedy is, rather, a tragedy of open access, in which no property-rights systems are in place. Of course, not all common-pool resources are governed in this way. Indeed, much of the work of those who George Caffentzis terms the "neo-Hardinians" has been focused on the question of how and where the "tragedy of the commons" arises, or might be averted, through detailed study of the attributes of common-pool resources that are more or less conducive to communal or individual ownership.[22]

A flourishing research agenda has thus emerged in the past few decades focusing on common-pool resource management regimes (successful and otherwise). It suggests that, subject to limitations, coordination and cooperation among users may evolve even without external incentives or state control and thus avoid "tragedies." In other words, under certain conditions, resource users are capable of sustainably managing the resource themselves. As Elinor Ostrom notes, "Examples exist of both successful and unsuccessful efforts to govern and manage common-pool resources by governments, communal groups, cooperatives, voluntary associations, and private individuals or firms."[23] This observation usefully reframes the issue as a problem of collective action in which emphasis is placed not on ownership but rather on the appropriate alignment of institutions and incentives enabling individuals sharing a commons to develop and sustain successful management models. When they are successful, a number of important "benefits of the commons" arise, including more effective management; better information (because it is rooted in local knowledge); equitable management (variously defined in terms of, for example, participation in decision making or sharing of resources); and lower management costs (because of volunteer and in-kind labor, or reduced regulatory requirements because of collective community scrutiny of resource use).

But sustainable common-property regime management, at least according to the scholarly literature, is only possible under a specific set of conditions. The academic evidence indicates that common-property regimes function well only in a small, defined set of cases because significant barriers exist to effective commons management. Under some conditions—including a small geographical area with well-defined boundaries, low levels of mobility, a small community with a high degree of social capital, and an overlap between residential and resource use location—common-pool resource water-management regimes have proven successful.[24] Common property water-management regimes, however, tend to perform poorly when any of the following conditions apply: a large number of users, uniformly abundant water resources, a large spatial scale with low population density, or a mismatch of jurisdictional and hydrological boundaries.[25] In short, although in some cases cooperative management institutions can function effectively to avoid depletion of common-pool resources such as water, there are also important limits to collective action.[26]

We can now begin to answer the question: Why do common-pool resource management regimes arise? One school of thought emphasizes the utilitarian aspect of these regimes as efficient systems for mitigating overconsumption among a delimited group of users.[27] A second school of thought emphasizes these regimes as ethnographic entities: they are historically and culturally embedded community arrangements, often linked to the dynamic defense of relative local autonomy and held together by moral economies and community bonds, which are adapted to the local environments in which they are rooted but also articulated within broader legal-political frameworks. A third, distinctly different school of thought—perhaps best characterized as the "communitarian commons"—presents idealized community constructs based on (at times essentialized) notions of solidarity and presumed equity.[28] Regardless of the reasons for their creation, observers agree that common-pool resource management systems are typically non-capitalist (although they may be market based and may be articulated with capitalist processes of economic exchange). They also appear to be quite resilient, even to the point of (in some cases) resisting external pressures to prohibit and/or reform systems of collective tenure and property rights.[29]

This is in part because a key element of sustainable common-pool resource management regimes, according to some scholars, is what Frances Cleaver terms a "moral ecological rationality," in which water users operate in a web of social relationships that support an ethic of environmental care. Ethnographic studies have demonstrated that water is part of a broader material-symbolic domain.[30] David Mosse, for example, argues that "water is not simply an exploitable productive resource or a physical input for agriculture. It is also a medium through which a variety of social relations have been structured.... Water systems are repositories of symbolic resources— they are a part of the symbolic production of locality."[31] In other words, the commons is not merely an instrumental resource-management tool undertaken by self-interested individuals; it is a web of relationships operating within (to use Michael Walzer's term) a "sphere of justice" with specific rules, norms, and moral commitments.

The "Commons" in Action: The Case of Bolivia

What are the implications of formalizing recognition of these commons management arrangements for water property rights? To answer this question, I return to the parable of Bolivia, below. But first it is important to note that the term "property rights" is highly elastic: it covers a broad array of use privileges, which are historically and geographically contingent. The variation in property rights for water around the world (and even within North America) is highly suggestive of the malleability of the bundle of entitlements that constitute water property rights in different places at different times. This comes as no surprise if we define property as a political relationship enforced by the state (following legal scholars) or as a social relation (in Marxist terms). Understood in this way, the political potential of property—to be regressive or progressive, to be more or less environmentally destructive—becomes quite evident. The conceptual point here is that property is not a fixed object framed by economic rationality; rather, it is a social relation that can evolve over time, in which different institutions and power relations are at work.

This perspective is helpful in understanding Bolivia's new irrigation water law, passed in 2004.[32] At one level, the new law is an example of the institutionalization of long-standing practices of commons water management in formal legal systems—promoted by some as a means of resistance to market environmentalist reforms introducing private-property rights and private sector participation in water supply. The Bolivian law followed years of mounting public protests against the introduction of market-based principles in water resources management, including attempts to introduce individual water property rights and, eventually, tradable water markets.[33] In reaction, citizens' groups argued that traditional, collective *usos y costumbres,* widely recognized throughout the Andes in the management of common property resources, were being displaced.[34] This was, opponents argued, a moral as well as a management issue, for *usos y costumbres* embodied water use practices well adapted to Bolivia's water-scarce, rigorous climate and to the indigenous cultural practices (both material and symbolic) predominant in rural areas and peri-urban zones.

The groundswell of public support within Bolivia after the Cochabamba and La Paz water wars, as well as protests against proposals for large-scale groundwater exports from Bolivia to a Chilean mining zone, lent impetus to the reforms demanded by irrigators: legal recognition of traditional uses, integrated water management predicated on multiple rather than single uses, and decentralized participatory water management.[35] The demands were supported by extensive research indicating that indigenous water-use practices were more efficient than the new proposals favored by the Bolivian government, which were based on granting concessions by unit of volume per time for specific uses. In the wake of the *guerra del agua,*

Bolivia's national government formed a Consejo Interinstitucional del Agua (CONIAG), mandated to draft a new water management law through consultation with a broad range of social actors—including those opposed to the water law in 2000. In fact, the government met these demands, and followed up with plans to develop a registry of traditional water rights and to reform early market-oriented water legislation, actively promoted by the newly elected government of Evo Morales, Bolivia's first president of indigenous origin, who had been involved in the Cochabamba water war.[36]

The resulting reforms to water legislation are a triumph if one considers that thirty-two previous attempts to implement new legislation had failed. Although still in its preliminary stages, Bolivia's water law reform is an example of the legal entrenchment of common-pool resource management regimes. This, at first glance, seems appropriate, particularly given the importance of traditional or indigenous cultural practices in Bolivia. Indeed, the recognition of *usos y costumbres* and the guarantees of irrigation rights to indigenous and farming communities were met with approval across the political spectrum within Bolivia.

But there are other, potentially less positive implications.[37] First, the formalization of these rights amounts to their privatization, insofar as private rights are allocated to individual rights holders (communities in this case). In other words, recognizing indigenous water rights, in this instance, implies removing water from the sphere of the "commons" (belonging to all Bolivians) and allocating it to private (community) owners. I realize that my characterization of the creation of collective water rights as "privatization" may seem objectionable to some. But it is technically correct: ownership passes from public (government) to private owners. Consider, moreover, the fact that the risk of exclusion is high for those who do not belong to an eligible community (for example, peri-urban residents, or rural residents who are not identified as "indigenous"). Second, the possibilities for the Bolivian state to plan or intervene in water resources management are potentially restricted. An ethic of exclusive ownership of water resources by some communities implies the difficulty in reallocating resources (for example, for drinking water to fulfill the state's new commitment to drinking water as a human right). Third, water rights create the possibility of the creation of water markets—in which water rights owners are to be reimbursed for allowing "their" water to be diverted to others. In other words, the danger here is that a system of water rights akin to private-property rights is being created. This is aligned with mainstream development models, which have evolved in an ideological environment in which government's ability to finance and support redistributive policies has been considerably reduced (hence the support of Bolivia's reforms by well-known proponents of market rights). But although these rights may provide some communities with an opportunity to engage in market exchange (e.g., via the sale of water rights, or the use thereof), they may not necessarily enable the poor within or beyond these communities to reap the benefits that markets can putatively provide; nor do they create the context in which coherent, integrated management of water resources can effectively be conducted.[38] My parable of Bolivian water management thus suggests some important concerns that might be raised about the use of legal norms of collective water rights as institutional mechanisms for formalizing "commons" management approaches.

CRITIQUING "WATER DEMOCRACY": QUERYING COMMUNITY-LED WATER SUPPLY MANAGEMENT

In urban and industrialized areas, particularly in wealthier countries, common-pool resource management regimes for water services have been frequently displaced by state and private sector provision (as explored in chapter 3). Particularly in cities, it is more usual to encounter community governance rather than outright ownership of water resources. In these cases, unlike a true commons, resource users are not always owners (or usufruct holders) of place-based water resources, but may own water-associated infrastructure or have been granted a degree of control over water governance through state-initiated consensus-based or participatory management practices. The division between "commons" and "community-based water management" in table 6.1 speaks to this distinction.

The past two decades have witnessed a resurgence of community-based, "delegated" or "collaborative" water management, which may be broadly defined as the involvement of nonstate actors in decision making for water management.[39] Similar to the broader trend toward "comanagement" of environmental resources, this frequently (but not always) implies the delegation of some degree of decision-making power to nonstate actors at lower scales of governance, and collaborative decision-making processes, often emphasizing consensus and trust building. Uniquely, however, delegated water management can also imply the use of a hydrographic boundary such as the watershed rather than political boundaries such as the municipality or region. Perhaps the most novel aspect of delegated water-governance partnerships is the involvement of a large number of stakeholders representing diverse interests who treat each other more or less as equals, guided by the principle that decision making should not be left solely to government experts. This increase in community involvement has occurred for a number of reasons, most importantly the putative shift from "government" to "governance" in which nongovernmental actors play a more significant role than in the past, thereby posing a challenge to conventional theories of governance in which the nation-state is the primary locus of political power and decision making.[40]

This shift has been associated with new approaches to participation in environmental governance and planning more generally, as well as increased acceptance of the legitimacy of community involvement in watershed-based management as a means of achieving political legitimacy for the integration of water-related goals. These goals have historically included such matters as the integration of land-use planning and water resources management at a local scale, particularly with respect to nonpoint source pollution, water quality management, coastal estuary protection, and protection of aquatic species. The trend is evident not only in a wide range of lower-income countries but also in high-income countries.[41] In the United States, for example, multistakeholder water-governance partnerships receive financial support from state agencies in several states; the growth of collaborative models of water governance, involving nonstate actors and multiple levels of government has been most notable in the western states, and has also been promoted by aid agencies and multilateral development banks as part of broader agendas of water sector reform.[42]

Community-based water management is thus far more comprehensive (in terms of integration of water management goals and spatial extent) and widespread (in terms of numbers of cases) than water cooperatives, which tend to be focused on the ownership and management of a single water provider—usually a water supplier ("retailer").[43] Nonetheless, water cooperatives merit examination, particularly because they have been attracting increasing attention as an alternative to conventional state or private-sector water-utility management options. Cooperatives are widespread in rural areas in both developed and developing countries, and in some cases (such as Finland) they are a predominant mode of water supply provision in both rural and urban areas.[44] Although size, scope, and degree of association with (or independence from) municipal governments vary substantially, cooperatives share in common a notion of "associative self-governance," in which consumers of essential services have greater control over service provision. These alternative ownership and management structures are intended, at least in theory, to resolve the trade-offs between shareholder and customer interests evident in the case of privatized (or, indeed, publicly run) monopoly services.[45]

Specific approaches vary among countries. In Britain, for example, associative self-governance has taken the form of "mutual" nonprofit corporations; in the United States and Canada, there are nonprofit corporations or
(largely in rural areas) water associations, particularly in the western states
and provinces.[46] Large-scale cooperatives, particularly in urban areas, are
rare; the largest urban water cooperative in the world (in Santa Cruz, Bolivia) serves a population of approximately seven hundred fifty thousand.[47]
But other, similar corporate structures exist, the largest probably being
the water supply utility that supplies 3 million customers in Wales, Dŵr
Cymru/Welsh Water, which was created as a nonprofit company owned by
its members and limited by guarantee (a classic management form for chari-
ties in Britain) in 2001.[48]

The success of the Welsh case suggests that, under the right conditions,
large-scale cooperatives can function effectively. This challenges the view
that cooperatives are best employed at a very small scale, in areas where
the state is unable, and the private sector uninterested, in service provision.
Aligning the incentives of customers and owners (by making owners members), reducing risk (and thereby the cost of capital and consumers' bills),
and creating efficiency incentives through the link between lower bills and
cost reduction (rather than lower costs and profit maximization) are some
of the key advantages of such a system. Cooperatives, and other similar
governance structures, may also be able to incorporate some of the recognized advantages of delegated water governance: access to "local" expertise that can improve the quality of decision making; the ability to adapt
regulatory programs to meet local conditions; empowerment of stakeholders (particularly those traditionally marginalized); reinforcement of "social
trust" between stakeholders and reduction of conflict over competing uses;
greater cooperation in information sharing; and greater political legitimacy
(and thus enforceability) of water management goals (such as water conservation or pollution control).

But there are also several critiques we might make of community water-
supply management. The first pertains to critiques of participation, in
which the involvement of community actors tends to legitimize policy and
environmental programs.[49] In poor neighborhoods of urban areas in developing countries, for example, participation may simply be a label for
strategies that seek to transfer costs from water companies to low-income
households—through, for example, voluntary labor for construction of networks.[50] A longer-term risk when communities are left to supply themselves,
as the Cochabamba case illustrates, is the entrenchment of two-tier services,
with poorer households forced to rely on services of inferior quality and

desirability. Participation may also be deployed as a technique of "normalization," intended to reform or even eradicate community water-use and management practices, particularly in contexts where indigenous communities continue to actively control local water sources.[51] And participation does not necessarily increase capacity.[52]

The rejection of state provision implicit in advocacy of community water management of urban water-supply systems is also problematic. The state remains, in many instances, the best vehicle through which consumers' interests can be balanced against one another, and against other interests. The need to balance equity and sustainability suggests the need for the continued, active role of the state in setting and enforcing water management criteria in community-managed initiatives. This is particularly the case in developing countries where the assumption that cooperatives will provide services in rural and peri-urban areas not served by municipal utilities will lead (and has led) to dual access standards and will foreclose the possibility of spatial and social cross-subsidies that have been widely used in some wealthy countries to support universal provision. It is important to qualify this argument, nevertheless, with the observation that both the state and the market at times undermine the ability of communities to engage in collective action in order to solve resource problems and govern resources. But certain strategies for recourse to the state can avoid the pitfalls of governance failure that I first addressed in chapter 1.

For these reasons, opponents of privatization often assert that governance models seeking to empower citizens are meaningful only when implemented in tandem with alternative service-delivery approaches via public-sector business models.[53] Customer corporations (in which incentives are structured toward cost minimization for a given service-quality level, rather than profit maximization for a given cost-minimization level) are an example.[54] Another example is corporatized public water companies (publicly owned, yet operating on modified commercial principles).[55] Customer-controlled models of utility regulation, in which consumers are not merely consulted but rather have formalized channels of input into decision making and some degree of decision-making power, are one example. Direct board representation (or even customer-dominated boards), external regulatory advisory bodies composed of consumers, or internal consumer-affairs bureaus are other examples.[56]

However, cooperatives and delegated water-governance models also have potential limitations. The consumer focus on local concerns (and, in particular, on minimizing bills) may reduce the degree to which cooperatives are willing to undertake costly long-term environmental sustainability initiatives (particularly at the watershed scale); indeed, this was one of the concerns that persuaded the Scottish government to create a public authority for water in Scotland rather than to mutualize.[57] Overall costs may be greater, and efficiency lower—particularly for smaller organizations without the requisite technical expertise.[58] An emphasis on consensus may lead to politically workable solutions rather than long-term economically and environmentally sustainable solutions, particularly in the case of unequal power relations and inequitable representation of consumers and other stakeholders in decision making (the "elite capture" phenomenon).[59] The large amount of (often volunteer) time required to maintain cooperatives raises questions about the long-term sustainability of consumer oversight, particularly in light of increased expectations on the part of governments and donors regarding participation—to the extent that it has been deemed the "new tyranny."[60]

In the wake of the partial retreat of private water multinationals, for example, advocates of public-private-community partnerships laud the actual and potential contribution of small-scale private enterprise and NGOs in service provision, particularly to the urban poor.[61] But the term "community partnerships" may be used as a euphemism for devolving water supply to informal providers and leaving poor, peri-urban, and rural communities to their own devices;[62] or for using volunteer community labor or concessionary finance as a means of subsidizing otherwise unprofitable private-sector water-management operations.[63] Others favor the "formalization," "regularization," or "legalization" of informal water suppliers.[64] Another contested example of this "residual commons" is the "public-public partnership," in which public water-supply utilities with expertise and resources are partnered with those needing assistance, often in smaller urban centers. Proponents have acknowledged the political pitfalls of promoting public-public partnerships in the wake of failed private-sector contracts, particularly the potential for such partnerships to be promoted as a strategy for less profitable communities, allowing more limited private-sector contracts to cherry-pick profitable communities and enabling international development assistance funds to continue to be directed toward private sector involvement.[65] We might term all of these phenomena the "residual commons" (table 6.2). Of course, the label "residual commons" implies a broader definition of the term commons than the neo-Hardinian version given above, suggesting a definition of the commons not solely with reference to common-pool resources but rather to a broader range of collectively created and shared resources.

TABLE 6.2. *Categorizing the Commons*

Types of commons	Example
Private club	Nineteenth-century clubs for social services (such as health) patronized by upper class in London, England
Commodity-producing commons (resource cartel)	Maine lobster workers
Subsistence-producing commons	Household gardens for self-provisioning—widespread in developing countries
Defensive commons	The public domain or a union: seeks to protect people or resources from the predations of capitalism
Inverse commons	Each user adds, rather than subtracts, to the total resource. E.g., users of the Linux operating system.
Reciprocal commons	A system of mutual aid, entitlement, and responsibility at a community level. E.g., water co-ops in rural areas across Canada.
Residual commons	Whatever is left after the claims of private corporations and the state have been satisfied. A good example would be the "public-private-community partnerships" in urban areas in the global South, where poor communities resort to commons-type arrangements for basic services (often in collaboration with NGOs), because nothing else is available.
Religious commons	Resources under Islam (which are the property of all of the faithful, not any state or set of nation-state citizens)

Source: Compiled in collaboration with the Forum on Privatization and the Public Domain.

In table 6.2, each type of commons operates along distinct axes of exclusion and inclusion. A "commodity-producing" commons, for example, functions as a resource cartel; only members of the commons have access to a specific resource (such as a fishery), which they may exploit as commodities, for profit, according to the rules agreed to and enforced by members. A "subsistence-producing commons," on the other hand, produces goods for consumption rather than for sale; household gardens for self-provisioning of necessary foodstuffs, common in Africa and throughout the developing world, are classic examples. Whereas the first type of commons is consistent with, and indeed fully inserted into, capitalist economies, the latter is more ambiguous; nonmarket production and exchange operates beyond the direct scope of markets but indirectly subsidizes the labor power of waged workers when and if they do enter the market.[66]

This conceptualization of multiple forms of collectively created resources enables us to see the various, and at times inequitable, forms of commons that may be created by communities, and provides one means of understanding how "community" solutions may reinforce an unequal (and unsatisfactory) status quo.

Table 6.2 also suggests the need to disrupt the opposition drawn (often implicitly) between "commodities" and "commons" in the water privatization debate. In fact, many commons are used to produce commodities. One might even argue that a corporation (characterized by internal cooperation and redistribution via worker ownership and share distributions) is a kind of capitalist commons. If we follow this logic, then the opposition between commons and commodities—and its conflation with the state (public)/market (private) binary—becomes untenable. Of course, opponents of privatization often deploy the concept of the commons as a means of asserting that collective management—whether by communities or the state—is not only preferable but necessary, and thereby argue that private ownership of water supply will invariably conflict with the public interest. Here, the commons serves as a metaphor of a sort of premarket state of grace, which market relations have wrongfully invaded (table 6.3).

Opponents of the commons fall into the same trap. For example, those who oppose commons management of water resources often assume that water can be divested from communities. Similarly, proponents of market approaches to water management argue that users can (and should) be redefined as individual customers or users with a merely utilitarian interest in water, rather than a collective of citizens alert to the religious, cultural, symbolic, and environmental dimensions of water. These latter arguments are misguided; on the contrary, work by anthropologists has shown the degree to which these noneconomic factors both shape water use and influence opinions about water privatization.[67]

TABLE 6.3. *The Commons versus Commodity Dichotomy*

	Commons	Commodity
Definition	Public good or trust	Economic good
Pricing	Free or "lifeline"	Full-cost pricing
Regulation	State-led	Market-led
Goals	Livelihoods and maintenance of resource base	Efficiency and profit
Access	Commons membership	Ability to pay
Manager	Community/state	Private companies/market

The commons/commodity binary is also misleading when applied to water because it obscures a historical reality: there exist more than two types of property relations for water. In Roman law, for example, we can identify at least four distinct property relations for water: *res nullius, res publica, res privata,* and *res communis omnium,* corresponding roughly to open access (such as the high seas); state (territorial seas, navigable rivers); individual (a riverbank or riverbed); and communal property (water in a flowing stream). Indeed, in countries with a legal tradition descended from Roman law, the concept of water as *res communis* still persists.[68] Customary rights, treaty rights, and aboriginal rights further complicate this picture. We misrepresent the historical evolution and contemporary diversity of water rights and water-use practices when we ignore these many types of rights, conflate public and communal rights, or focus solely on public versus private property rights, as the commons/commodity binary encourages us to do.

THE LIMITS OF COMMUNITY

Let me sum up the critiques made above. First, proponents of commons and community water-supply management often essentialize the concept of community. This glosses over the "partial citizenship"[69] that many residents of developing countries experience on a daily basis, in which claims on the state must be constantly reiterated and negotiated and are by no means assured. Admittedly, much of the literature on collective, community-based forms of water supply management tends to romanticize communities as coherent, relatively equitable social structures, and to overlook inequitable power relations and resource allocation practices.[70]

Indeed, common-pool management regimes and community water-supply management are not necessarily equitable or democratic.[71] The fact that these regimes are not equitable should not be taken to imply, however, that they are not functional. On the contrary, in many instances, inequitable commons arrangements can be highly functional, in terms of goals such as conservation of resource stocks or a "floor" (or minimum threshold) for resource access for the poor. Academic scholars of the commons have long recognized this point, but in popular debate the ideal of the commons is often conflated with notions of equity and democracy.[72] Free-market environmentalist proponents of the commons, for example, often idealize community resource management as an autarkic mode of regulation free from state involvement. Others frame the commons as a downscaled version of the state, a miniature polis. Neither of these views is commensurable with the reality of the commons, understood as a means of collectively managing common-pool resources.

Second, and related to the previous point, commons and communities can be exclusive and regressive, as well as inclusive and progressive.[73] Appeals to the commons run the risk of conveniently overlooking the fact that the commons are a mechanism to "privatize solidarity"; common-pool resource management regimes are, in fact, defined by exclusion (i.e., closed membership that limits access to resources by nonmembers). Nor are commons or community management systems necessarily equitable. This is the case, for example, with respect to urban water supply systems in middle-class areas, which can be as exclusionary of the poor as conventional systems.[74]

A third set of issues pertains to the broader political-economic landscape in which commons or community solutions are pursued. Advocates of community involvement often fail to account for the broader political-economic dynamics that foster desires on the part of elites for greater community involvement. For example, an endorsement of community involvement risks condoning the cherry-picking of profitable or otherwise attractive cities, neighborhoods, and regions, by both public and private water-supply utilities, with some communities (as in the Cochabamba case) left to their own devices. An argument in favor of the commons thus risks reinforcing the post–Washington Consensus, where states, markets, and communities play complementary but definitely unequal roles. This is particularly the case when community water provision occurs by neglect rather than design— the default option for those whom neither governments nor markets are able or willing to serve. Here, the role of the state in encouraging redistributive models of resource management, progressive social relations, and redistribution is more positive than those making calls for a "return to the commons" would perhaps admit.

A fourth objection pertains to the mistaken assumption, prevalent in the community water supply literature, that implementing the correct organizational form (such as a cooperative or community ownership) will automatically give rise to desired changes in behavior and thus in management outcomes.[75] Organizational form can prohibit certain actions (such as the sale of assets to investors), but it will not automatically result in changes in behavior, although it can create conditions in which these changes might be possible. For example, improving governance through involving consumers in decision making can improve transparency and accountability but can rarely deal effectively with issues of financing, access, and operational management.[76] Simply put, the imposition of public or community control is not a sufficient condition for better management.[77]

Nor should we be confident that organizational models can be easily transposed from one locale to another: the application of rural commons-management models to urban areas, for example, is by no means straightforward. For these reasons, opponents of privatization often assert that governance models that seek to empower citizens are meaningful only when implemented in tandem with alternative service-delivery approaches.[78] Customer corporations (in which incentives are structured toward cost minimization for a given service quality level rather than profit maximization for a given cost-minimization level) are one example;[79] another is corporatized public water companies (publicly owned, yet operating on modified commercial principles).[80]

Indeed, the intensity of the debate over privatization has tended to obscure the diversity of social and economic relations through which water supply is regulated around the world. If we look closely at specific cases, we invariably find apparent contradictions regarding governance—that is, the process through which stakeholders provide input, decisions are made, and decision makers are held accountable. For instance, consider the close regulatory oversight and political pressure applied to British water-supply utilities, which had supposedly transcended political "interference" by becoming privatized.[81] Compare this to the seeming unresponsiveness of public water authorities to the needs of poor families without household connections, even in the face of the supposed greater accountability of public ownership and management.[82] This example suggests that the question of access depends not on public versus private ownership but rather on political leverage that different groups can bring to bear on actors in donor agencies, NGOs, and (most importantly) governments. Admittedly, this viewpoint rather simplistically pits civil society against political society (to use Chatterjee's terms), but it captures the "partial citizenship" that many residents of developing countries experience on a daily basis;[83] that is, claims on the state must be constantly reiterated and negotiated, and they are by no means assured.[84] Demands for the extension of water supply to poor households must be seen, from this perspective, as attempts to deepen democracy by making claims over material emblems of citizenship—the basic public services necessary, particularly in urban areas, for dignified human lives.

As explored above, examples abound of this "deep democracy" in action.[85] But it is important to challenge the presupposition that "deeper" democracy automatically implies greater efficiency and effectiveness in the management of local services. For example, the degree of power that consumers can effectively assert does not, of course, flow automatically from consumer participation in utility regulation. We should be equally cautious of the "local trap," in which proponents of community or commons water management conflate effective democratic representation with "lower" scales of governance (for example, a community or watershed).[86] The local scale, according to its proponents, is more responsive to community needs, more empowering, more effective, and more cost-efficient than higher scales of governance.[87] We can, of course, challenge this uncritical acceptance of the rhetoric of the local.[88] Brown and Purcell (2005), for example, articulate the dangers of the "local trap": the assumption that organization, policies and action at the local scale are inherently more likely to have desired social and ecological effects than activities organized at other scales.[89] On the contrary, one can equally argue that communities, particularly if they are poor, are likely to lack the requisite capacities and institutions. Moreover, common-property regimes do not necessarily have to be applied at the local scale; we should not conflate "commons" with "communities." Common-property regimes may be as large as a nation-state (as with the welfare state) or a region (such as worker cooperatives in the Basque country).[90]

The constraints imposed by the biophysicality of resources such as water further complicate this appeal to the local. In the case of water, appeals to the commons often equate the local scale with the watershed as the "natural" scale of organization for ecological systems and human activities.[91] From an ecological perspective, the watershed is not necessarily the ideal scale. It does not correspond with biomes, or even groundwater distribution through aquifers. For individual users, such as farmers, the watershed may be so large as to be relatively meaningless.[92] But the watershed scale is also too small to deal with the issues raised by the fact that water is a flow resource—implying that the actions of one user can affect others beyond the watershed. The flow of water through the hydrological cycle implies the need for scales of governance both "higher" and "lower" than that of the watershed. Moreover, the idealization of the watershed may simply endorse the "downshifting" of responsibility, without critically examining whether local communities have the resources and capacity necessary to govern, or whether local governance translates into equitable and meaningful participation, significant influence over decision making, and accountability or empowerment.[93] In the case of water, for these reasons, we should design nested scales of governance—rather than (or in addition to) a local commons. In short, we should reject the assumption that communities can solve all water supply–related issues, particularly in urban areas. Improving governance through involving consumers in decision making can improve transparency and accountability, although it is not always sufficient to deal effectively with issues of financing, equitable access, and ecological sustainability.

CONCLUSIONS: CONSTRUCTIVE AMBIGUITY

Appeals to the commons and community are highly ambiguous. As with other such terms of broad, even universal appeal—democracy, justice, freedom—this ambiguity gives these terms enormous political reach, while leaving much room for quibbling over interpretation. As I have argued above, appeals to the water commons within the debate over water privatization are often predicated on misinterpretations and misrepresentations of community water supply and the commons. In practice, commons and community water-management approaches also have important limits. This means that alternative water-management strategies predicated on concepts of the "commons" or "community," although they might in theory provide us with progressive options for environmental management, are in practice often highly constrained and compromised, as demonstrated by the Bolivian case.

What, then, should we think about the degree to which "commons" represent a progressive alternative to conventional approaches to water supply management—stalemated between proponents of privatization and defenders of classic models of government provision? In an attempt to introduce some precision into the debate, I have argued that the notion of the water commons, in current practice, is more relevant to and widespread in rural areas, where community ownership and collective property rights in water— particularly in regions where water is vital for subsistence—are central to community life. These water collectives are enormously varied, and not necessarily fully equitable (although this is not necessary, the academic literature on the commons emphasizes efficient, self-sustaining systems). But they often represent highly adapted, time-tested solutions to the problem of accessing water in difficult ecological (and indeed social) conditions. Here, granting collective property rights to water can, if appropriately handled, be progressive; for it constrains the possibilities for encroachment by both state and market actors. But the case of Bolivia suggests that water rights do not necessarily have progressive (in the sense of redistributive) outcomes. This failing arises in part because the commons requires exclusion, as well as inclusion. Accordingly, attempting to advocate the commons as a blanket model for social organization (for example, through arguing that all environmental resources are commons) both dilutes and misrepresents the term. It also blinds us to the potentially negative implications of its implementation.

The situation is more complex in urban areas or in situations where water supply has been industrialized through large-scale hydraulic networks. The transfer of "commons" management institutions from rural to urban areas is, I have argued, fraught with difficulty. Community governance (i.e., involvement in decision making) is more common than ownership, although it may be far from universal. In this latter case, claims for rights tend to focus on the human right to drinking water rather than on collective property rights in water. In the case of modern water-supply networks in urban areas, because of the high degree of technical complexity and scale of hydraulic networks at play, governance (i.e., participation in decision making) is the key mechanism by which communities exercise influence.

Let me reiterate this point about the undeniable necessity of government responsibility for water supply. Recourse to the state is necessary if we are to supply urban water on an equitable and universal basis to all. To demand less, or to demand that communities take on the responsibility for water supply themselves, risks condoning rather than addressing governance failure. It also runs the risk of devolving the costs of water supply to the poor, and of creating entrenched and inequitable two-tier systems. Proponents of community solutions need to avoid, in other words, endorsing divestment by the state (and governing elites) of responsibility for urban water-supply services—whether couched in the language of "decentralization," "participation," or "community management."

Some might contend that, on the basis of this reasoning, we should dismiss notions of commons and community. Certainly, these approaches are an uneasy fit with the technocratic expertise that governs the day-to-day business of managing large-scale technical systems such as water supply networks. But dismissing the notion of community altogether would be both unrealistic and misguided. We should not dismiss community-management strategies merely because of their complexity and limitations. Nor should we summarily dismiss their claims to the moral dimension of property rights, in which the nature of water precludes or constrains private ownership, usually on spiritual or environmental grounds, and requires forms of trusteeship or stewardship. The moral and symbolic dimensions of water are important reasons why communities have a critical role to play in water supply management. But it is precisely these dimensions of water supply that create the multiple, locally rooted community practices that are often an uneasy fit with market logic and state-management mechanisms. In short, definitions of community are slippery and multiple, and they disrupt our normative distinctions between "markets" and the "public sphere."

Rather than rejecting (or attempting to erase) this diversity, then, we should recognize and embrace the resulting ambiguity as constructive. To the extent that this endorses pluralistic approaches to water governance and multiple, community-driven approaches to water management, this is to be welcomed. In other words, proponents of community control are correct in arguing that conventional models of public and private sector management do not exhaust the range of alternatives to be considered in the water supply sector—although they are incorrect when they argue that commons and communities are antitheses to the market or governments. We should not, therefore, entirely abandon the notion of the commons. But we should be cautious about its progressive potential, particularly with respect to the reconciliation of environmental and social-justice concerns when the commons is applied as a management option. It is to this set of conceptual and practical issues that chapter 7 now turns.

7

POLITICS AND BIOPOLITICS

DEBATING ECOLOGICAL GOVERNANCE

Water is both political and biopolitical. Flowing through the hydrological cycle, water links individual bodies through the cycling of waters from one organism and one ecosystem to another. As it cycles, water transgresses geopolitical boundaries, defies jurisdictions, pits upstream against downstream users, and creates competition and conflict over its uses as a source (or input to processes of modernization, industrialization, urbanization) and sink (for the disposal of what are colloquially known as the "effluents of affluence"). Water is thus intensely political in a conventional sense: it is implicated in contested relationships of power and authority.

But water is also biopolitical: carrying with it vectors of disease and pollution, water simultaneously connects individual bodies to the collective body politic. The regulation and control of water-borne bodily wastes, the disposal of which has become an intensely private activity under modernity, is an inescapably collective act, essential to the health of the population as well as the individual. Water is thus biopolitical in the Foucauldian sense: modern governments seek to optimize both water resources and individual water-use practices in order to secure the health and productivity of the population. This control is enacted through formal regulation, but it is also self-policed through a normalizing cultural aesthetics of health and hygiene, from water bodies to individual human bodies.

Scholars have developed these insights in studies of the propagation of modern hygiene practices in colonial contexts. In Indonesia, for example, discourses of water-related hygiene were deployed both as evidence of the superiority of "modern" colonists over "natives," but also to justify racially segregated water-supply systems.[1] Similarly in Australia, public-health practices were allied with often overtly racist strategies of population management; controlling Australia's Aboriginal population served to demarcate and define "white" behavior and identities.[2] This was not unique to colonial contexts: Jean-Pierre Goubert has shown how the class identity

of the urban French bourgeoisie was constructed in part through the spread
of the private boudoir and the use of increasingly elaborate (and expensive)
hygiene-related devices and practices, which in turn justified increased investment in networked water-supply systems to "water" the city.[3] Similar
processes have been documented for other cities, such as New York and
Athens, suggesting that the way we use and relate to water is quintessentially biopolitical: simultaneously fetishized and self-consciously constitutive of both individual identity and population health.[4]

In physically connecting individuals to the collective, and in linking humans so integrally to the nonhuman world, water poses the problem of
collective action in a particularly acute way. Various disciplines have attempted to formulate terms with which to describe this interconnectedness.
Economists and social scientists speak of the health benefits of clean water
as a "public good": nonrivalrous (in that one person's enjoyment of water-
related health does not detract from another's), nonexcludable (in that individuals cannot be excluded from the collective benefits of public health),
and nonsubstitutable. Natural scientists speak of water's circulation via the
hydrological cycle and have coined the term "ecohydrology" to describe the
interactions between water and the environment. Political ecologists prefer
the term "hydro-social" cycle, expanding this set of relationships to include
the built (or "artificial") as well as "natural" environments, emphasizing the
fact that water connects individuals not only materially but also politically
and socially. Religious traditions have a completely distinct set of terms,
revolving around water's symbolic importance and sacred character: water
is a metaphor for spiritual purity, the vehicle for sacraments and ritual puri-
fications, and an object of worship. All of these perspectives frame water as
an integrator of our social, spiritual, political, and ecological selves.

The problem of collective action posed by water is, in other words, not
simply confined to the social sphere. The "collective" includes humans and
the ecosystems to which we belong. This is equally true for all other resources; humans are material beings, and our relationship with the environment is inevitably *metabolic,* in the Marxian sense.

This, in turn, implies the need to adopt a relational formulation to the
question of water supply. On the one hand, we need to take account of water
as a flow resource: water supply is merely one aspect of the hydrological
(or even hydro-social) cycle. Viewed in this fashion, the glaring absence of
considerations of sanitation within the privatization debate (and indeed this
book) becomes more obvious and more troubling, but also perhaps more
amenable to resolution. We need to view water, in other words, in *ecological*
terms, in addition to seeing it as a resource input to water supply systems.

On the other hand, a relational approach suggests that water is articulated with other social relations such as property: land use and water use are inextricably interlinked, in both cities and rural areas. As earlier chapters explored, the link between people and urban property (e.g., via tenure) is central to explaining urban water-use practices. This implies that an explanation of the social construction of urban water-use practices must articulate the relationship between land and water, and between the social relations that govern these resources (such as land tenure and water rights).

A relational approach thus poses the challenge of balancing materialist and social constructionist perspectives, as much of the literature on "social nature" (whether urban or not) attempts to do.[5] Why is this useful? I argue, below, that this allows us to think through the ideological commitments of the industrial model of water development first introduced in chapter 1. From this perspective, the problem of urban water supply can be reframed: it is not merely a problem of land tenure, or poverty, or managerial capacity, but also a set of problems stemming from the industrialized model of water on which the "municipal hydraulic" paradigm is predicated. Simply put, as we industrialize water systems, we create conditions that degrade water quality and threaten water security: "governance failure" is thus ecological and technical (in Ellul's sense of the term "technique") as much as it is institutional.

Below, I frame these arguments in terms of a concept of ecological governance. I argue that this concept might provide the basis for productive strategies of dealing with governance failures for urban water supply. In part, this is because the former concept reframes the latter as an ecological as well as a sociopolitical phenomenon. Taking ecological governance seriously, in other words, implies thinking about privatization as an environmental as well as a political-economic process (similar to my arguments about urbanization as an ecological process in chapter 4). This, in turn, may help us rethink the tensions identified in earlier chapters—between land and water, property rights and human rights—and thus complement the notions of community and the human right to water explored there.

Before I go further, let me anticipate one likely objection by asking the question: Are we right to worry about environmental issues when over one billion people around the world are without access to sufficient supplies of safe water? To answer this question, we might refer to the debate over poverty and environmentalism. One influential strand of thinking about the relationship between environmental concerns and economic growth asserts that individuals' desire to protect the environment emerges only after basic needs are satisfied and a certain level of societal wealth is attained.

Following this line of thinking, we might be tempted to conclude that environmental issues are a distraction in the water privatization debate. But this would be misguided, because environmental concerns are, in fact, central to the livelihoods of the urban poor. Poor environmental quality is costly, in both health and economic terms. This is as true for lack of access to safe drinking water as it is for many other issues. And water carries an additional set of threats for the urban poor, who live within the interstices of the city (in floodplains and along riverbanks, on steeply eroded slopes and marshy land). In the zones where the poor settle, water is often a threat to physical safety, both in terms of flooding and poor water quality. Better water governance—addressing ecological concerns across urban watersheds—would do much to address these issues. Attention to environment-related water concerns is thus an imperative, and not a luxury, for the urban poor.[6]

The broader debate over the legitimacy and centrality of environmental issues in the provision of public services is the backdrop for this chapter, which situates questions over privatization and urban water supply within discussions of the ecological dimensions of urban water governance. Why is this focus merited? In previous chapters, I have concluded that neither government nor private provision provides a viable answer to the urban water crisis. The most widely promoted alternatives—the human right to water and various forms of community water-supply management—address redistributive concerns in economic and social terms. But they overlook important environmental aspects of the urban water crisis: water scarcity, threats to water quality, and ecological sustainability. This view demands that we carefully examine claims to a human right to water and to the ubiquity and necessity of community water management. Let me be clear: as argued earlier, the human right to water is morally defensible and tactically useful in the search for more equitable water-supply access. And communities undoubtedly have an important role to play in water management. But we risk committing both social and ecological injustices if we do not situate these strategies within a broader framework of ecological governance.

How might we begin to grapple with these issues? I begin, below, by suggesting a framework through which we might analyze the ecological dimensions of privatization. It is based on a political-ecological (rather than political-economic) perspective on urban water.[7] This discussion leads, in turn, to a reconsideration of the paradigm of distributive justice on which debates over water privatization are usually based (and on which the human right to water is predicated). This opens up the possibility, in the final section of the chapter, for recasting questions of community in light of recent debates over ecological governance. I argue that this, in turn, might help us solve some of the conundrums we face in our conventional approaches to the roles of states, markets, and communities in water management.

What do I mean by the term "ecological governance"? This framework (and there are many others like it, with different labels) is based on two central claims.[8] First, human activity should be understood as occurring within, and as an integral part of, ecosystems. Second, recognizing the interrelationship between humans and the environment requires that we evolve alternative, ecologically sensitive systems of governance.[9] These ideas intersect in suggestive ways, as we shall see, with the expanded notions of governance and community explored in previous chapters.

THE ENVIRONMENTAL DIMENSIONS OF PRIVATIZATION

To begin, let us consider the environmental dimensions of privatization by summarizing the dramatic changes in the relationship between markets and nature that have taken place over the past two decades. Markets have encroached in an unprecedented manner on previously uncommodified aspects of the natural world, from genetic resources to atmospheric carbon. Our responses to environmental crisis have increasingly been mediated by the doctrine of neoliberalism (introduced in chapter 1), "the doctrine that market exchange is an ethic in itself, capable of acting as a guide for all human action."[10]

The result is what Stephen Bernstein terms "liberal environmentalism," which he defines as the belief in the "compatibility of environmental concern, economic growth, the basic tenets of a market economy, and a liberal international order."[11] Simply put, liberal environmentalism is a set of strategies for environmental management in which market principles are entrenched in the doctrine of sustainable development. The rise of market-based regulatory mechanisms and private sector participation is not, in other words, unique to the water sector: this is a broad-based trend across resources and environmental services.

Why has this trend occurred? Opinions differ, as captured in the distinctions among the phrases "liberal environmentalism," "green neoliberalism," and "natural capitalism." From a political-economic perspective, liberal environmentalism has emerged in response to serious environmental threats to profitability under capitalism, namely, the depletion of the environment's ability to act as both a source (of commodities) and a sink (for the disposal of wastes and pollution). James O'Connor terms this threat to economic growth the "second contradiction of capitalism": the creation of environmental conditions that undermine continued capital accumulation.[12] Liberal environmentalism represents a response to this contradiction, in which capitalism "involutes" (to use Cindi Katz's term) as it seeks to turn these threats into new opportunities for capital accumulation.[13] This creates new opportunities for technological innovation, substitution, and private appropriation. Two examples related to water supply are the growth in consumption of bottled water (one of the fastest-growing segments of the beverage industry) in response to concerns over declining water quality; and the emergence of water-quality treatment technologies and businesses focused on pollution cleanup.

Of course, there is nothing inherently new about this trend: we need to remember that the "environment" (as we conventionally term it) has always been enrolled in capitalism. For example, nineteenth-century political economists recognized the important role played by the enclosure of the commons in creating both a proletarianized workforce and new commodities from which profit could be extracted.[14] Their insight—and in particular that of Marx—was that capitalism, like all human economic systems, is predicated on our metabolism of nature, which we both literally and figuratively ingest and transform, altering ourselves in the process. Under capitalism, no aspect of the environment is left untouched, as science and technology are mobilized to commodify and transform our social and natural worlds. From this perspective, all of capitalism might be termed liberal environmentalism.

Yet the specific incarnation of liberal environmentalism occurring in the contemporary period is distinctly different. It is transgressive, particularly when it enables the creation of new techno-natural objects through genetic engineering: biotechnology is the classic example, with Dupont's OncoMouse as its emblem. Solicited by capitalist markets and enabled by technoscience, liberal environmentalism undermines a foundational myth of modernity: the distinction between the natural and the social. Scholars of political ecology, human ecology, and science and technology studies have used various terms to describe the leap of knowledge necessitated by this perspective: Neil Smith's and Bill McKibben's arguments about the "end of (first) nature," Bruno Latour's "hybrids," Erik Swyngedouw's "socio-nature," and Sheila Jasanoff's "co-production" of science and society all seek to dispense with the nature-society dualism so central to modern thought.[15] These views are both reminiscent and suspicious of environmentalists' calls to view humans as part of nature rather than distinct from it.

In Ulrich Beck's formulation, these responses might be characterized as the negotiation of the risks posed by environmental "bads" (rather than goods), spurred by the fear of their effects that comes to dominate collective politics and individual psyches in industrialized societies.[16] In urban areas in so-called developing countries, for example, sanitation and refuse are environmental "bads," the spatial distribution and economic costs of which are hotly contested. The reaction to the encroachment of markets thus takes on a qualitatively new character. The threat posed by capitalist modernity is not merely to an existing social order; it is, rather, a deeply existential threat to the basis of life itself. In making the connection between psyche and society, Beck's formulation of the risk society suggests a broadening of Karl Polanyi's concept of the "double movement"—in which capitalism oscillates between dis-embedding and re-embedding economic activity within society.[17] Indeed, an underlying theme in the environmental consciousness that came to the fore in social movements of many kinds in the second half of the twentieth century is the re-embedding of our societies and economies within the ecological lifeworlds of which we are a part. If liberal environmentalism dis-embeds the economy not just from society but also from the socioecologies on which we all depend, then appropriately restraining technoscience, reinventing capitalism, and reimagining our worldviews of socionature are tasks that gain a new urgency.

The Ecological Impacts and Drivers of Privatization

These arguments are provocative and open to dispute. The extent to which capitalism is capable of "greening" itself is, for example, a subject of active debate. These arguments, although important, are not central to my analysis here. The framework of liberal environmentalism is useful for our purposes because it allows us to frame the ecological dimensions of water privatization; both driven and justified by deterioration in environmental conditions, water privatization has the potential to result in significant impacts on environmental quality.

The environmental impacts of water privatization, I should emphasize, are not necessarily negative. As I have argued elsewhere, we must recognize the potential of water sector reforms, allied with liberal environmentalism, to improve environmental quality.[18] Where water is discursively recoded as part of the environment (rather than a mere resource) and this new definition is formally incorporated into enforceable (and enforced) regulatory frameworks, the environment may come to be represented as a legitimate water "user" whose interests are to be balanced with—or even prioritized above—those of consumers. To give a concrete example: greater legitimacy and protection for an expanded "environment" have been embedded in water regulation and management in England and Wales, and impressive improvements in water quality have come about as a result of the revised regulatory framework introduced in 1989 in tandem with privatization (largely in response to EU legislative requirements). Let me be clear: my argument is not that privatization is causally related to improvements in environmental quality. Rather, my point is that privatization (and liberal environmentalism more generally) is constituted by (and constitutive of) processes of reregulation that may result in improvements in environmental quality. The difference between these two arguments is subtle, but important. The former asserts a causal relationship; the latter, in contrast, cautions analysts not to jump to conclusions about causality, and in particular, not to assume that environmental quality can only decline under water sector reforms.[19]

With this in mind, we can reconsider the arguments made for and against water privatization, focusing on ecological concerns. Proponents of liberal environmentalism argue that through establishing private-property rights, employing markets as allocation mechanisms, treating environmental goods as economic goods, and incorporating environmental externalities through pricing, environmental goods will be more efficiently allocated. Concerns over environmental degradation and the inefficient use of resources will, they argue, be simultaneously addressed if individual initiative and private sector savvy displace government inertia and incompetence, and if market-led technological innovation is permitted to substitute or otherwise compensate for depleted resources. This provides context for the broad range of water sector reforms (introduced in chapter 1) that tend to accompany water privatization.

Privatization opponents object that these water sector reforms are merely a veil for an unprecedented intensification of capitalist plunder of the environment; some would add that markets are, in many cases, the *cause* of our environmental woes, and certainly no solution. The nature of their objections varies. Political economists argue that the profit motive may result in waste rather than stimulating efficiency, particularly where resources are abundant. They note that some innovations—particularly those beneficial to the environment—are so costly or difficult to implement that government action is required. Moreover, they emphasize that resources are often characterized by "market failures": effects generated by market activity that are not priced and therefore not accounted for by economic actors. To the extent that these "failures" are irremediable, acceptance of the market-failure argument implies the need for government intervention.

Environmentalists and some ecological economists raise additional objections. For example, perfect substitutability is unattainable; natural limits— whether to growth, to the ability to absorb wastes, or to rates of exploitation—cannot be transcended by markets any more than by governments.

From this perspective, conservation and protection, rather than efficient allocation, should be our primary concern.

In lower- and middle-income countries, where many liberal-environmentalist initiatives were pioneered over the past three decades, other objections are often raised. Many believe that liberal environmentalism is part of a broader process of the global expansion of capitalism into the periphery, in which privatization plays a key role in creating "new realms for proletarianization and private appropriation of public property."[20] Simply put, water privatization is an example of the encroachment of private control over environmental resources. This argument is often combined with the assertion that the deployment of markets is misguided because resources are not merely economic goods but have spiritual and ecological dimensions central to the subsistence livelihoods of communities. "Global justice" (or "alter-globalization") activists sometimes add a further objection, that private-property rights are the problem (rather than the solution): the environment is a "commons," which should be collectively owned on a not-for-profit basis. From this perspective, private-property rights are objectionable, because private, for-profit ownership creates ecological harm and fosters social disunity. This reasoning implies a rejection of liberal environmentalism in favor of an expansion of commons or community management. In turn, these utopian views imagine a revival and expansion of the noncapitalist economic systems still prevalent in some areas of the world. Vandana Shiva is one example: an ecofeminist from India, Shiva advocates that water management be framed in terms of ecological justice, grounded in spiritual as well as material connections with the environment, and in traditions of water democracy in southern India.[21] This perspective—in a pattern typical of views emanating from the "global South"—links an ecological worldview explicitly with political and resource management norms. For a "Northern" perspective, we might refer to the work of Spaniard Pedro Arrojo, who advocates for a "new culture of water" that dissociates water management from its traditional sources of expertise (engineering, economics), and proposes, instead, an ethic of water governance predicated on the recognition of ecosystem integrity and the right of communities to ecological, as well as hydrological, sustainability.[22]

Reframing Water in Ecological Terms: Neither Public nor Private

These fictional utopias of a "new culture of water" have much in common with proposals for Integrated Water Resources Management. Both sets of approaches suggest that recognition of the cyclical nature of water as a flow resource—extending beyond urban boundaries and transcending geo-political jurisdictions—must be the primary basis for water management. Without such an approach, say proponents, we will be unable to manage water sustainably. Protests against privatization often articulate a similar viewpoint: protests against a given technological approach (large dams and centralized hydraulic systems) are often allied with protests against a specific socioeconomic worldview (of commercialized management of natural resources, driven by the imperatives of economic growth).

Questions about the relative merits of public and private management need to be situated, from this perspective, within a broader understanding of ecology and hydrology (the study of the biophysical dimensions of water as a flow resource). Here, it is important to note that water presents managers with at least three seemingly intractable dilemmas: dealing with competition between multiple users of water resources; balancing the multiple scales at which water is managed; and responding to the mismatch between geopolitical and administrative boundaries, on the one hand, and hydrological boundaries on the other. Water poses these dilemmas in part because it is a flow resource that constantly transgresses political boundaries. Impacts of water use—both positive and negative—are felt far downstream, in other jurisdictions. Yet water is used locally: cheap to store but heavy and thus costly to transport, water is most often used close to the point of abstraction.

As a result, water creates a tension between the local scale and higher scales of governance and is characterized by competition not only between multiple uses but also frequently between upstream and downstream users. As a flow resource, water is, moreover, the supreme integrator; given water's ability to dilute and transport pollutants (sometimes over long time scales), the nature of the impacts of water use by one user on another is often difficult to discern. Dealing with competition and multiple jurisdictional scales is made more difficult by the fact that most jurisdictional boundaries are oblivious to watersheds. The integrated management of water raises the question of *which* scale is the most ecologically meaningful one: the river basin, in the case of surface water sources that have undergone significant human interventions; the groundwater aquifer, in the case of communities reliant on wells for drinking; or the local stream, in the case of agricultural water users. But these do not necessarily coincide with ecologically meaningful boundaries.

This framing provides insight into potential solutions to a particularly intractable question within privatization debates: the issue of scale, so central to the tension between community and corporate systems. Water is biophysically multiscalar, but it is used and disposed of locally. This suggests that we might frame our thinking about alternative governance in terms of approaches that accommodate (while not blindly mimicking) this multiscalar nature of water.

Recognizing water's ecological functions also requires that we abandon the argument, embedded in privatization debates, over whether water is a public or private good. Its status is irrevocably ambiguous. To be sure, water supply has many aspects of an economic good, and has (or generates) an economic value in some of its competing uses. Yet competing views of water—based on spiritual, ethical, or environmental criteria—continue to be voiced in opposition to this notion. These are predicated on water's ecological and spiritual functions, but also on the recognition that the complex terrain of water and land rights, overlaid with water use practices, forecloses overly simplistic distinctions. As explored in earlier chapters, this is most obvious in urban areas of lower- and middle-income countries, where public and private sector provision of water supply typically overlap—and often compete. But it is also true in high-income countries. This suggests that we look elsewhere for definitions of the identity of water. For example, one potential alternative is the public-trust doctrine: that specific resources cannot be truly owned by anyone, that they are to be preserved for public use, and that government action is required in order to maintain these resources for the public's reasonable use.[23]

The foregoing analysis suggests that debates over privatization of urban water are simply too narrow, for several reasons. First, they focus on public versus private to the exclusion of community. Second, they focus on service delivery while overlooking environmental issues (such as "in-stream flows"). Third, urban water issues cannot be neatly separated from the broader watershed context in which they are situated. Fourth, urban water issues are inseparable from questions of urban land rights (this is equally true in rural areas). Hence, questions of urban water supply must also engage with issues of water resources and nonurban users within the watershed. A grasp of the co-construction (and interdependence) of urban and rural socioeconomies via the mobilization of water is necessary to understanding contemporary urban water access, scarcity, and conflicts. Maria Kaika has demonstrated, for example, the interrelationship between the Greek state's drive for modernization, the construction of the Athens water supply system (predicated on twinned discourses of the threats of scarcity and national backwardness), and the resulting hydraulic conquest in the city's hinterland, allied with programs of "improvement" that aligned rural socioeconomies more closely with the city's needs.[24]

This approach encourages us to recognize the multiple networks of actors in urban water supply, extending well beyond the "formal" water supply system and beyond the city's boundaries. This leads to my fifth point: that our frame of reference in governance debates should not be restricted to the network and its users but should, rather, include a larger, integrated set of technological systems and ecosystems, which emerge in response to urbanization as an ecological (and material) as well as socioeconomic and political process. In taking on these expanded notions of governance and urbanization, we should not abandon the social-welfare considerations that typically dominate the privatization debate, but should balance them with environmental limits and strive for the need for inclusive decision making about how to address these considerations while respecting these limits. Finally, as explored below, this implies the need to rethink our conventional approach to redistributive justice.

RETHINKING DISTRIBUTIVE JUSTICE

Much of the water privatization debate and studies of the impacts of privatization are grounded in a modern, Western framework of distributive justice.[25] In other words, debates over the respective roles of states and markets are often framed in terms of the assumption that the criteria for justice are primarily related to material goods, and that our "sphere of justice" (to use Walzer's term) encompasses only individual humans. This underpins conventional formulations of human rights, including the right to drinking water.

At least three objections can be raised here. First, one might argue that the ultimate goal of water supply management decisions should not simply be the allocation of water as a *material* good. Amartya Sen's concept of capabilities is one way of formulating this approach: the equal moral worth of all human beings creates an obligation to assist all individuals to develop their capabilities (or human "potentials," in Kantian terms) through both material goods *and* requisite social institutions. From Sen's perspective, material goods are a means, but not the goal. This leads us to a definition of justice as the maximization of everyone's potential, which requires a concept of solidarity and collective action. Access to water supply should be a *means* and not an end; and it is predicated on some notion of solidarity as well as redistribution.

Second, we might go further still and reject the ecological blindness of theories of distributive justice that implicitly assume unlimited material wealth, as captured in the alter-globalization (originally Gandhian) mantra "Enough for everyone's need, not for everyone's greed." This is congruent with a broad array of sustainability arguments that critique modern, Western notions of development and progress (particularly those that are predicated on an assumption of limitless—or at least perfectly substitutable—resources). These sustainability perspectives suggest, equally, a critique of the logic of the developmental state, in which redistribution of material goods is a central justification for the extent, and indeed existence, of government. Simply put, distributing water as a material good risks failing to meet criteria of ecological sustainability, and it distorts our views of the purposes and role of the state.

Third, one might also argue that our emphasis on *distribution* is misplaced. In this, I echo Marxian arguments, which suggest that it is misguided to consider the treatment of economic (re)distribution separately from issues of economic production. Here, one might argue that we should focus our attention on what Marxians often term the *means of reproduction* (rather than solely production) in a broad ecological sense. By this, I intend to invoke an approach to water that is explicitly ecocentric, bringing to mind notions of stewardship (rather than exploitation) and acknowledging the embedding of human societies within ecosystems. This perspective enlarges our critique from the social to the ecological aspects of the distributive justice paradigm. Another way of framing this point is to suggest that, rather than debating the role of states and markets in distributing resources, we should turn our attention to the ways in which we can collectively act as stewards of the socioecological lifeworlds of which we are a part.

Let me give a simple example of how we might rethink our approach to distributive justice and reframe our concern with the impacts of water management as an issue of solidarity, linked to issues of social and ecological justice. My example pertains to tariffs for urban water supply. Three sets of approaches are usually advocated, each of which reflects the limits to conventional approaches.

First, the typical set of water sector reforms advocated by promarket policymakers would impose full-cost-recovery tariffs. Only this approach (they would argue) would enable the water supply utility to operate on a financially sound basis and recover its costs. This is necessary if the water utility is to sustain the revenue flows needed both for supporting the costs of infrastructure maintenance and for supporting the costs of financing the inevitable capital expenditures required to finance network extensions and new water-resources developments. The impacts of pricing on affordability and access might be of concern, but these should not be addressed within the water sector (but rather, for example, within the welfare system).

Second, and in contrast, those with a socially minded agenda would propose, instead, that tariffs be linked to some index of income (such as property value or house size) to ensure affordability (typically defined as a threshold of average income). Or they might declare in favor of rising (or "block") tariffs, pegged to volumetric usage, which would penalize higher consumption levels while transferring subsidies to lower-volume users. This would imply cross-subsidies within classes of consumers. But if a large proportion of consumers were poor, the water utility might not recover its costs. Accordingly, some argue that water should be free. In all cases, subsidies from higher orders of government, or external aid agencies, would likely be necessary. The impacts of such a strategy on the water utility might be of concern, but this should not preclude satisfying the human right to drinking water as the first priority.

Those with an environmental agenda would adopt a third and different position, likely arguing in favor of "environmental economic" pricing. Here, cost recovery is still a goal, but it refers to an enlarged definition of costs, including those borne by the environment owing to water abstraction, consumption, and disposal. This proposal implies even higher costs, posing potentially greater problems of affordability, which might only be partially offset by the rising block tariffs that an environmental perspective would favor as a means of reducing water consumption. Their recommendations would meet with stiff resistance. Those prioritizing the human right to drinking water would reject the increase in prices (and decline in affordability) that the environment "imposes" on individuals. Those prioritizing the financial sustainability of water utilities would worry about the impacts of reductions in consumption on water utilities' abilities to continue financing their functions.

These three positions (which respectively privilege economic, social, and environmental considerations) are thus, apparently, mutually exclusive and irreconcilable. If we attempt to simultaneously consider social- and ecological-justice concerns, however, we realize the contradictions between these positions are potentially reconcilable. Water, like energy, is an unusual commodity because it has a derived demand. It is used directly for drinking, but it is also converted in appliances (dishwashers, washing machines, radiators, toilets) into the desired outcome (e.g., clean dishes, clean clothes, warm homes, and sewage disposal). Whenever water is used, the conversion efficiency of appliances will vary, so issues of capital and running cost are involved.

It is true that raising water prices to encourage more efficient water use may also have the effect of cutting consumption of water services, as greater water efficiency requires capital expenditure that lower-income families may not be able to afford, because poor households typically (where they have them at all) have lower-efficiency appliances. So the immediate effect of a water price rise will be a drop in consumption, particularly by the poorest families, unless accompanied by the necessary capital expenditure in the home. And a drop in consumption may have dangerous effects—such as the cholera epidemic in South Africa (discussed in chapter 5) that followed the introduction of water pricing. Supporting people's capabilities—to an adequate level of personal health, for example—is predicated on sufficient amounts of safe water.

But here we have only an *apparent* contradiction between social and ecological sustainability. The solution is widely deployed in the energy sector, where policies of providing subsidies for energy-efficient appliances in the home—advocated as simultaneously sustainable and equitable—have been implemented to address "fuel poverty," and where the limited nature of full-cost tariffs, by themselves, in achieving conservation goals is widely accepted.[26] Similar "water-poverty" policies provide subsidies to increase water efficiency in homes, targeted in particular at groups for whom water expenditure is above a threshold percentage of income. This avoids the regressive effects of the commercialization of water, while still supporting goals of financial and ecological sustainability. It implies a notion of solidarity that goes beyond mere cross-subsidization of tariffs: the financing (whether through user levies or transfers from government) of household water infrastructure. And this notion of solidarity is not merely restricted to humans: the preservation of ecosystem health is a justification for this type of "water-poverty" approach.

Producing Scarcity

The preceding example connects to a point made in earlier chapters (particularly in chapter 4) regarding water scarcity. In chapter 4, I argued that urban water scarcity is "socially produced" (that is, mediated by human actions, both discursive and biophysical). In other words, water is mediated by social as well as infrastructural and biophysical processes. As it circulates through the hydro-social cycle—a complex network of rivers, streams, aquifers, pipes, water law, meters, quality standards, garden hoses, consumers, and leaking taps—water is mediated not only by rainfall and runoff but also by infrastructure.

Moreover, water is not bounded solely by the physical infrastructure that abstracts, treats, and distributes it and removes wastewater; the flow of water depends on demand patterns, customers' expectations about water quality and pressure, and policies and laws. Water circulation, in short, is dependent on institutions and practices as much as on the hydrological cycle. The fixed capital embodied in the material network of mains, pipes, pumps, reservoirs, and sewers can thus be viewed as an artifact of the hydro-social cycle, shaped by successive generations of social, technical, and economic practices. This fact is often obscured with respect to water, which becomes naturalized (or fetishized) as water supply becomes the norm, and as infrastructure is buried beneath city streets.[27]

This social construction of scarcity is replicated across the globe. Twenty-first-century urban dwellers consume enormous amounts of water (although consumption of water by domestic users is only 4 percent of total water consumption, the largest component of which goes to the agricultural sector). Each inhabitant of London uses perhaps 160 liters per person per day, and the inhabitants of Los Angeles closer to four times that amount, and demand per capita increased steadily throughout the twentieth century.

This pattern of highly unequal access is reproduced at the global scale. Whereas wealthy residents of desert states such as Nevada consume on average over 950 liters per person per day,[28] the World Health Organization estimates that 1.1 billion people worldwide do not have access to safe drinking water, and that 2.4 billion are without access to adequate sanitation.[29] The amount of water mobilized by human societies increased dramatically over the twentieth century; indeed, there are few resources we mobilize in such large volumes, and yet adequate supplies of sufficiently clean water remain scarce.

Social constructionism roots the causes for this scarcity partially in the realm of human choice and thus suggests the remedies are open to collective action. It does not deny the material scarcity of water but rather suggests that our experience of material scarcity is socially mediated. This implies, in turn, an important question: How might we move beyond scarcity?[30] If we are to reformulate distributive justice in environmental terms, it is this sort of question, rather than a narrower debate over "public versus private," that, one might argue, should be at the forefront of academic and public debate.

Toward Ecological Governance

The preceding discussion asks whether we should reject some or all of the premises of the modern, Western notion of distributive justice in expanding the terms of debate over water privatization. These arguments might be perceived as suggesting a sort of Third Way argument, in which states, markets, and communities play complementary roles in achieving shared environmental ends. There is, of course, nothing novel about this latter view. Traditionally the debate about states, markets, and communities has been framed as a choice between alternative economic systems, and the environmental blindness of this approach has been the focus of sustained critique. Nonetheless, the point remains valid: privatization is, in part, an environmental phenomenon, and any solution to water supply debates must take environmental issues into account.

This has implications for the ways in which we approach water: as a resource, an input to public health, an ecological service, or a spiritual entity? Many of these views imply a rejection of the conceptual commitments underlying the "municipal hydraulic paradigm." As an example, we might refer to Shiva's perspective (discussed above), which reflects the Frankfurt School's rejection of technocratic systems of rationality that accompany Western modernity—without rejecting modernity's promise of better lives for all.[31] Others, such as legal scholar and Cochabamba "water-war" activist Rocío Bustamante, emphasize solidarity rather than spirituality as the grounding principle for collective action, but this, too, can extend to incorporate environmental concerns. As Bustamante argues, in her call for a reframing of both water rights *and* the human right to water:

> The basic idea is to rethink the concept of rights in relation to the management of natural resources such as water, and to disrupt these concepts in order to foster a more creative relationship with nature not rooted in the idea of rights, but rather in the idea that natural resources are there for all of us to share, including non-human beings. We must start with principles such as solidarity, which transcends the idea of a "right" because a "right" always requires that we identify who is entitled to a given "right," who is supposed to guarantee it, and who has legal standing. By contrast, if natural resources are thought of as a commons, they do not belong to anyone in particular.[32]

These are obviously utopian views. A precondition for their realization is some mechanism of solidarity (typically moral or spiritual), guided by conscious choice (rather than human nature). A second assumption is the indivisibility of nature from society (or "socionature"). This ecocentric view demands that we extend solidarity to nonhumans. The third and perhaps more subtle dimension of these views is the notion of "coproduction," which asserts that nature and society are mutually constituted. In other words, the interrelationship between our technologies, cultures, economies, and polities and our lived environments is productive. It leads us to ask about the relationship between the technological and socioeconomic arrangements for exploiting resources and our political commitments.

Here, I am in agreement with Shiva's and Bustamante's framing of an ecocentric view that extends solidarity to nonhumans. At the same time, however, it is important to be cautious about utopian ecological views. We should be wary of the implications of the critique that Shiva and Bustamante apply to common-property rights, and in particular the implicit notion that we might abolish property rights—individual or collective—to water. As I argued in earlier chapters, the human right to water *and* community water rights represent important, although imperfect, means of articulating aspirations for equitable access to water supply and resources. The human right to water and common-property rights to water are, of course, neither necessarily equitable nor ecologically sustainable, but their outright rejection is not the answer. Rather, collective action for water stewardship can exist only in an enabling institutional environment that includes individual rights and collective property rights, in an environment of mutual checks and balances between governments and communities. The eco-utopian commons is no more of a solution, in my opinion, than the concept of the "global commons" critiqued in chapter 6.

How, then, are we to think about the ecological dimensions of alternatives to privatization? There is no single answer to this question. Our relationship with water is quintessentially local and place based. But we can suggest some principles that might guide our discussions. We could begin with a sense of the sacredness of life and acknowledge the relationship between human and ecological health. This in turn implies the need for some form of collective, rather than individual, stewardship. There are precedents for the formalization of such a principle in law, such as the doctrine of "public trust." This, in turn, is incompatible with the maximization of consumption or profit characteristic of our current systems; it implies, rather, the (re-) embedding of water management within social and ecological relationships. This would require, in turn, that communities (at various scales) be viewed as the ultimate locus of legitimacy in decision making, giving them radically democratic, collective control over infrastructure networks and the ecosystems in which they are embedded. This does not, I should emphasize, imply a "small is beautiful" mentality, or an automatic delegation of power to the local scale; as with other environmental issues, some water issues are national, international, or even global in scope.

We should also be wary of autarkic, utopian depictions of community that suggest that freedom from government oversight (or from interaction with the market) is desirable, or even possible. Communities, and any common-pool management regimes or community management processes with which they are associated, are not antitheses to, nor hybrids of, markets or governments. In the contemporary political-economic conjuncture, they are dialectically related to states and markets. This suggests that community engagement should be predicated on "solidarity without romanticism," as Rutgerd Boelens puts it.[33]

In determining how communities might engage in specific aspects of water management and governance, we would apply a principle of subsidiarity to both governance and technological systems, devolving management and decision making to the lowest appropriate scale. Inclusion would be built via webs of relationships reinforced by reciprocal responsibilities and entitlements. In doing so, we would emphasize equality (although not necessarily equity) in redistribution. In many (although not all) cases, this would be the scale of local communities (whether neighborhoods, parishes, or watersheds). Adopting an ecocentric view, communities would invoke systems of mutual solidarity, but not restrain solidarity to humans. In acknowledging the relationship between land use and water, we would place constraints—sometimes serious ones—on our choices about the spatial organization of human activities. We would view scarcity as a symptom of imbalance between different socionatures, rather than as an obstacle to be eliminated or an inevitability to be tolerated. This would in turn support precautionary management to create the conditions for socially constructing abundance: enough water for everyone's needs (but not everyone's greed). In doing so, we would necessarily move beyond privatization, but also beyond many of the assumptions and commitments inherent in our current approach to water governance.

Corporate versus Community Control

We might also deploy the concept of community to usefully disrupt the government–private sector binary around which mainstream water policy revolves. But, as we have already discovered in earlier chapters, definitions of community are slippery, and the boundaries between community, government, and the private sector are not always clear. Perhaps a more useful distinction, for our purposes, is that between corporate and community control. The foregoing distinction between corporate and community control might be criticized as falling into the trap of technological determinism and oversimplifying the distinctions between corporate and community control.[34] But we can avoid the trap of assuming that certain types of resource exploitation are *invariably* associated with specific socioeconomic systems if we pay close attention to the social construction of specific types of technological systems, social organizations, and the scales at which they operate. And the distinction between "corporate" and "community," it should be emphasized, is best viewed as a continuum rather than a binary.

With these caveats in mind, we can revisit the distinction between corporate and community control first discussed in chapter 1. On the one hand, small-scale water-supply systems are more often directly controlled by communities. These tend to be in rural or peri-urban areas, and in the former, often deal with water resources (the engineer's "raw" water) for multiple purposes, such as irrigation, flood control, electricity generation, and household uses. These systems are often community built and used solely by community members. They consequently require a very different set of skills—highly adapted to their local environments and related to community norms of sharing labor and water—associated with local ecological and hydrological knowledge. Here, we cannot really say that water has been industrialized; its production is largely undertaken by artisanal (or, as they are known in development circles, "appropriate") technologies. As with corporate-controlled water supply, community control often functions well over long time periods, but it can also be inequitable and unsustainable. Here, the "human right to water" has less validity, and community "water rights" (with all of their pitfalls) greater applicability.

On the other hand, large-scale hydraulic networks tend to be controlled by corporate, bureaucratic structures, which can mobilize the necessary technical expertise and financing. Water supply, in this case, has been industrialized. In urban areas, where water supply is largely concerned with drinking water, this makes sense, because of their large scale, and because the production and disposal of drinking water, as a nonsubstitutable resource essential for life, have important public health and environmental dimensions. Corporate control is allied with what Ellul terms "technique": the set of rationalities and practices that emerge from, and in turn support, certain technological practices. Of course, all societies confront the simultaneous need to mobilize water resources for a range of uses, to mediate conflicts between users and uses, to build and maintain infrastructure, and to dispose of wastes while mediating jurisdictional scales. With large-scale hydraulic networks, the involvement of communities in water governance is necessary if these tensions are to be mediated (although they are unlikely ever to be fully resolved). But some degree of corporate control, and collective representation mechanisms that transcend individual communities, are also necessary. Here, recourse to the state (however imperfect) is imperative. Here, the "human right to water" (with all of its limitations) is an obvious means of recourse if corporate control of water supply, mediated by the state, fails to be equitable and sustainable. In other words, when we are dealing with urban water supply, the role of the state in encouraging redistributive models of resource management is more positive (and hence necessary) than those making calls for community water management would perhaps admit.

Nonetheless, in rural areas in many regions, common property regimes remain vitally important in water management. And in urban areas, the notion of the commons is useful because it demands that we question the inviolability and desirability of the property relations that underpin our norms of justice and individual rights. In querying property rights, the commons perspective invokes notions of justice as a social relation, invested with a moral dimension (rather than the modern view of justice as procedural). These ideas echo premodern thinkers, ranging from Plato to Rousseau and the Christian aesthetes, who argued that private property leads to disunity (for moral or political reasons) and that communal property is more conducive to social harmony. Advocates of the commons are thus often suspicious of the individualism central to modern, Western notions of justice. The critique that they make is thus analogous to the critique of human rights discussed in chapter 5.

Simply put, the commons as metaphor usefully invokes the problem of collective action and the issues of social and ecological justice with which it is intertwined. In its evocation of community control, notions of the commons provide us with "fictional" rather than "blueprint" utopias. As such, they are an exercise in expanding our imaginations, in the hopes of indirectly contributing to the practical task of constructing more equitable political ecologies, the specifics of which will vary among peoples and places over time. While it is unrealistic to assume that commons models can be replicated out of their social, technological, and ecological contexts, a clearheaded examination of how commons work in practice is useful in the task of constructing solidarity without romanticism. Accordingly, if we do want to evolve local, sustainable community water-supply systems, one challenge will be to leave our utopian views of community behind and to engage with collective water-management approaches—with all of their qualities and flaws, equities and inequities.

Concluding Reflections: Rethinking Governance Failure

What are the implications of this admittedly abstract discussion for the urban poor without safe water access? In previous chapters, I have contended that a human right to water, and the involvement of communities in water supply governance, are important elements of sustainable alternatives. But in this chapter, I have argued that we must go further, and rethink the conventional approach to water provision that guides both government and private provision.

My reasoning followed two main lines of thought. First, if we broaden our approach to redistributive justice to include the environment, the necessity of improving watershed governance becomes an imperative for both ecological and socioeconomic justice; protecting hydrological environments will also improve living conditions for the urban poor. For example, the need to address sanitation concerns—so often overlooked in the water supply privatization debate—becomes unavoidable. Moreover, the narrow basis on which the impacts of privatization are often evaluated becomes untenable. This, in turn, lends further weight to critics of water privatization; if evaluated on environmental terms (even if restricted to the question of sanitation), the flaws of the privatization agenda of the 1990s become even more apparent, and the need for alternatives even more pressing.

Second, attempting to embed ecological concerns within governance debates leads us to adjudicate various water management alternatives on the basis of both ecological and social considerations. On the one hand, the nature of water as a flow resource suggests that conventional frameworks of property rights and human rights are ill suited to governing water. This reinforces points made in earlier chapters about the limits of a rights-based approach. On the other hand, the nature of water as the "supreme integrator" lends support to collective ownership and management strategies. Here, the issue of scale is central. Water is incessantly on the move, linking global to local scales as it flows through the hydrological cycle. But, as we have seen, its use and disposal are localized, place-based activities; in urban areas, these necessarily occur at a larger scale than in rural communities.

Accordingly, if we follow the dictum that governance must be adapted to ecological conditions, the challenge made in earlier chapters to the public-private binary is reinforced. Governance failure is simultaneously ecological, socioeconomic, and political. On social grounds, the case for decentralization (or delegation, as it is sometimes termed) might seem compelling; but on environmental grounds, it is not. Indeed, multilevel governance is unavoidable given the nature of water as a flow resource, and governments are best able to carry out the coordination required. This conclusion does not imply that communities should play no role in water governance. On the contrary, the tension between the need for centralized oversight and community control (the "corporate-community" tension identified in earlier chapters) is inherent in our relationship with water, and I do not mean to suggest that it can be transcended. Rather, a consideration of environmental issues leads us to reinforce the critique (first made in chapter 6) of utopian representations of autarkic communities as the solution to the world's water problems. It also enables us to deliver a more comprehensive critique of the neoliberal imaginary (of autarkic corporate, individual, or community action in the context of government deregulation) that underpins the reform of water supply management along market lines.

In closing, I should add that these arguments may be new to readers largely concerned with privatization as a political-economic phenomenon. But they will be familiar to scholars and activists from the global South, where livelihoods, ecological issues, and social justice are intertwined concerns. My call to move "beyond privatization" through grappling with ecological questions is thus not merely an academic exercise. It is also an appeal to reorient our debates in ways that are more reflective of the concerns of those they are intended to serve.

CONCLUSION

BEYOND PRIVATIZATION

The coastal Brazilian town of Porto Alegre is perhaps best known as host to the World Social Forum, the annual gathering of global-justice organizations and activists. Few know that it is also exceptional for its approach to water supply management: over 99 percent of the city's population (of approximately 1.5 million) are connected to the water supply network, and nearly everyone pays their bills—both phenomena a rarity in Latin American cities.[1]

By most accounts, the city's water department (DMAE) is well run: it appears to be efficient, equitable, and effectively managed. Tariffs are socially and environmentally progressive: wealthy and high-volume consumers pay higher rates per unit volume, cross-subsidizing poorer households while encouraging water conservation. And water is nonetheless affordable: DMAE's bills are some of the cheapest in Brazil. Like other city departments, DMAE engages in "deep democracy": it is advised by a "deliberative council" that includes citizens, overseen by a "social audit" process as a citizen-oversight mechanism, and answerable to citizen preferences through "participatory budgeting" for improvement and expansion of services. And although DMAE is government owned, it outsources (contracts out) 40 percent of its activities to the local private sector.[2]

Before these policies were put in place, water was supplied only to the wealthy downtown. Following the election of the left-wing Partido dos Trabalhadores (Workers' Party, or PT) in 1989, the city administration's commitment to expanding water supply meant that new connections kept up with the rapid rate of urban growth, improving environmental water quality while making Porto Alegre immune to Brazil's recent cholera outbreaks. Although many cities in Brazil have experimented with privatization (often under pressure from national government and international donors), Porto Alegre has resolutely maintained its pragmatic model, which appears to have combined many of the best attributes of government, community, and the private sector. This is in line with its approach to public services more generally. And it is an approach that has paid off: the city has Brazil's highest Human Development Index score, and indeed scores at a level comparable to the capital cities of developed countries.

Curiously, Porto Alegre's water supply receives relatively little attention in academic and media analyses of water, and other success stories of effectively run public water supply systems—from Phnom Penh (Cambodia) to Bogotá (Colombia)—are similarly neglected. Media attention is often focused on so-called water wars and confrontation over the management of local water-supply systems by multinational water services firms (usually European or U.S.). And policy and scholarly attention is often focused on the litany of failures that characterize the world's urban water crisis. But this evades compelling questions: Why is Porto Alegre different? And can it be replicated?

Porto Alegre evokes an elusive goal of much water policy debate: urban water supply that is (relatively speaking) environmentally sustainable, socially acceptable, and economically viable. It is also emblematic of key arguments made throughout this book. Porto Alegre's experience suggests that the complex reality of urban water-supply access refutes conventional notions of "public" versus "private." It also suggests that the urban water crisis is not solely due to the misdeeds of government or the private sector but also to failures of governance and collective action. This, in turn, speaks to more general debates about deliberative democracy; the respective roles of states, markets, and communities in economic life; our collective response to environmental crises; and the inequities embedded in our conventional approaches to adjudicating questions of social and ecological justice.

These debates form the conceptual backdrop to my analysis, highlights of which I will briefly summarize here. Water privatization has been promoted as an urgently necessary solution to the world's water crisis, which we can briefly characterize as a lack of basic services, on the one hand, and degradation of water resources, on the other. The provision of water supply services by private, for-profit corporations increased rapidly around the world from the 1990s onward. Some of these companies were among the world's largest multinationals, epitomizing the rapid growth of private, for-profit activity in sectors previously dominated by governments. In lower- and middle-income countries, this trend was actively supported by influential actors such as the World Bank and the International Monetary Fund. The struggles among these corporations, development organizations, and NGOs opposing water privatization epitomized the antiglobalization protests of the late 1990s.

The resulting debate over water privatization has pitted those who argue that privatization is a panacea for the ills that plague the water sector—poor government management, growing water scarcity and pollution, underpriced and undervalued water resources—against those who reject any private role in water supply. Privatization proponents—which include private water companies, the major development banks, some bilateral aid organizations, and a range of proprivate NGOs—argue that water is an economic good and should be priced accordingly in order to ensure conservation. On this basis, so goes the argument, private companies will provide the improved performance and efficient management desperately needed to address the world's growing water crisis. Opponents, on the other hand, argue that this rhetoric is merely a veil for the theft of water from the public domain. Antiprivatization campaigners, including unions, environmental groups, religious groups, and a broad range of civil-society organizations, insist that water is a human right, a public trust, and a commons, the preservation of which is an imperative and profit from which is inherently unethical.

How has this book sought to intervene in this debate? My primary goal (as the title of this chapter suggests) has been to move "beyond privatization" through reconsidering the terms of the public-versus-private debate and constructively critiquing proposed alternatives to privatization.

THE PRIVATIZATION DEBATE

In chapter 1, I introduced a central dilemma indicative of the urban water-supply crisis: whereas many governments have consistently failed to adequately deliver water supply, private companies have typically failed to improve the situation. This dilemma is suggestive of a stalemate at the heart of our current debate on the roles of states, markets, and communities in the provision of environmental services. On the one hand, those loosely gathered under the banner of the post–Washington Consensus (or self-styled "neo-Keynesians") argue that markets and states should play complementary roles, with civil society (communities and NGOs) filling in the gaps. Governments should leave as much scope for market mechanisms as reasonably possible and restrict their role to that of umpire, regulator, and facilitator. Within this framework of government oversight, market mechanisms can achieve our desired environmental and economic goals. In order for this to occur, we need to "get the institutions right": hence the focus of public-policy debate in recent years on concepts such as social capital, good governance, and institutional economics. We might term this set of ideas "neoliberal institutionalism." On the other hand, we have witnessed a proliferation of proposals for a revival of the commons and community economies, concepts that have become increasingly central to alter-globalization (or "global justice") movements.

In subsequent chapters, I examined these points of view. Successive chapters sought to bridge debates over development and access to public services (particularly on the part of the poor in developing countries) and debates over sustainable management of the environment. Issues of governance, power, and politics were central to this analysis; they are often overlooked by academic and policy literatures focused on the relative merits of ideal-type public-versus-private delivery models.

My analysis dispenses with some myths widely propagated in water privatization debates: that water is a "global commons" (it is not); that the private sector wants to take over the world's water supply (it does not, or at least not all of it); or that scarcity has created a "global water crisis" (the converse is true: it is the crisis in our water management models that has created scarcity). This provides impetus for constructive criticism of conventional management models, as well as the most politically popular alternatives (the human right to drinking water, and community water management in various forms).

With this as my conceptual starting point, my intervention into this debate was framed around three questions. First, why has privatization emerged as an increasingly widespread mode of resource management, and what are the arguments made by proponents and opponents? Second, can privatization fulfill proponents' expectations (particularly with respect to water supply for the urban poor in developing countries)? Third, given that there are limits to privatization (particularly with respect to urban water supply to the poor), what might be the alternatives?

In answering these questions, I have proposed that although they differ in key aspects, models of government and private provision share an important similarity: their commitment to a modernist, humanist worldview of water. This worldview defines water as a resource, to be put to instrumental use by humans, via centralized, standardized hydraulic technology, in a drive for maximization (whether of water supply or profits), on the basis of a hierarchical management structure predicated on technical expertise, which creates an atomized relationship between individual users and the network. The blind spots of both sides of the debate are thus also similar: a relative neglect of water's ecological functions; a dismissive view of artisanal water harvesting and use practices; and suspicion (if not outright rejection) of place-based water knowledge and the spiritual dimensions of water. This state of affairs, I should emphasize, is congruent with the modernist approach to resources more generally, in which the rational, scientific management of resources is able to wreak technical miracles, but not without enacting a degree of ecological and cultural violence. As documented in chapter 2, the resulting problems were a central driver for the resurgence of water privatization in the late twentieth century. But, as argued in chapter 3, private sector management (and indeed ownership) of water cannot resolve these issues.

Water privatization is, in other words, not a panacea. As documented in chapter 3, private companies have retreated from the water supply sector in lower-income countries. As a result, the debate over alternatives is flourishing, particularly in lower- and middle-income countries (sometimes, imprecisely and imperfectly, referred to collectively as the global South). Many of the "failures" we observe in urban water supply are, in fact, common to the public and private sectors, in particular, an inability to adequately address environmental concerns and a systematic pattern of failing to extend water supply networks to the urban poor. For these reasons, private sector provision, particularly in the forms adopted in the 1990s, will not succeed in supplying water to the urban poor on a global scale. But nor do conventional approaches to government provision provide us with adequate solutions.

Problems with water access, pollution, and control have existed throughout recorded history. But, as environmental historian Ramachandra Guha has explained, the collective consciousness of environmental threats on a global scale—to water and a host of other resources—is historically unprecedented.[3] Indeed, the urgency of this debate is often framed in terms of the global water crisis, in which a significant proportion of people in developing countries (over 1 billion people, according to most estimates) are without access to sufficient amounts of clean, safe water on a daily basis. Framed in this way, the urban water-supply crisis raises the question about the limits—philosophical and political, discursive and economic, cognitive and material—of our models of resource exploitation and our instrumentalist approach to nature, of which privatization is only one element.

This framework suggests that the urban water-supply crisis is simultaneously socioeconomic and ecological: a global pattern of localized crises in which pollution, overexploitation, and poor governance have resulted in increasing water scarcity. These crises are simultaneously discursive, social, and material: the social construction and political mediation of a very real set of deteriorating environmental conditions. In practical terms, it means that (in many cases) access to water is a problem of distribution rather than absolute availability. This perception should give us hope. It suggests that scarcity is socially constructed: in many instances, sufficient volumes of water are available but are inappropriately exploited, wasted, and distributed. It also suggests that through conscious choice (and contestation), redistributive measures will allow us to go beyond scarcity and to enable water access for all. This is not to suggest that absolute water scarcity does not exist: in some places, this is a real threat. Nonetheless, the experience of scarcity, as argued in chapter 4, is mediated by social and political factors: simply put, our current approach to urban water governance "produces thirst."

BEYOND PRIVATIZATION: THE SEARCH FOR ALTERNATIVES

Accordingly, a recurrent theme of this book has been that sustainable solutions to the urban water crisis require us to move beyond the rather stale positions staked out in the public-versus-private debate. Three of the most popular alternatives—the human right to water, water management by the commons or the community, and ecological water governance—were the subject of successive chapters in part 2.

Human Rights versus Water Rights

As the chapters progressed, I asserted that access to water is properly viewed as a material emblem of citizenship: a symbol of political inclusion. This has much in common with the views of the proponents of the human right to drinking water. But I also argued that the many limitations of human rights (investigated in chapter 5) suggest caution. Nonetheless, we should not reject the human right to water altogether (as some suggest). Instead, we should perhaps best view human rights not as a *goal* but rather as a *tactic* in creating the conditions for more equitable allocation of water infrastructures and more democratic control of the ecosystems in which they are embedded. However, the limitations on this strategy should be emphasized. I argued that the notion of social rights (or the "right to have rights") is potentially more powerful and effective than a narrow focus on human rights. Moreover, the human right to water focuses only on drinking water and not on inherent aspects of rights to water resources: rights to water governance, rights to water resource use (complicated by their ties to land use and ownership), not to mention the contents (we might term this the "institutional architecture") of these rights. This suggests that we can not consider the question of the human right to water in isolation: the human right to drinking water and community water rights must be considered in tandem.

Community water rights, in other words, are complementary to—and in some cases necessary for—the realization of the human right to water. I have argued that the recourse to community (although at times the only default option available when weak or incompetent local states do not deliver) is also a means of evading state responsibility, while (when private firms are involved) turning a public-health catastrophe into an opportunity for profit. In cases of this sort, a human-rights based rhetoric can often enable communities to effectively articulate claims for basic government services and to demand accountability and responsiveness to community needs. Human rights are also a useful means of formulating critiques of the negative effects on the poor of the commercialization of services—whether carried out by the state or the private sector.

This strategy implies, as I explored in chapters 5 and 6, the articulation of human rights, property rights, and social citizenship rights; or, more specifically, the articulation of the human right to water, land tenure rights, and what Mitchell terms the "right to the city" (including public services associated with urban environments). This interrelationship between land and water issues has, of course, long been recognized in water management circles—via the framework of Integrated Water Resources Management. But rarely does this framework treat the articulation of land and water in urban areas as a socioeconomic and political issue as well as an ecological one.

Ecological Governance: Debating the Commons

But I have also argued that water's ecological functions—which render it inescapably collective—need to be taken into account. This suggests that forms of trusteeship or environmental stewardship (and the approach to property rights that this implies) are a potentially useful way of resolving key tensions between corporate and community control, between the individual right to drinking water and broader ecological concerns, and between the multiple scales at which water flows and thus must be governed. Human rights, in other words, must be allied with reforms to property rights and broader governance norms that enable the treatment of nonhuman water users as (at least to a degree) political subjects. The anthropocentrism of the human right to water will otherwise raise a specter of environmental harm. For example, what are the implications of calling, as some activists do, for water "for everyone, everywhere, forever?"[4]

If we do need reform to property rights, then where should we look? As explored in chapter 6, the most widely debated set of reforms to property rights within the current water privatization debate centers around the commons. Here, though, we are presented with a dilemma. On the one hand, we have the academic "neo-Hardinian" view of the commons as a narrowly defined set of common-pool resources, the management of which is amenable to improvement through better governance and optimized institutions and incentives. This view is congruent with weak sustainability perspectives that narrowly frame the environment as a set of environmental services, including water resources available for exploitation. The anthropocentrism of this interpretation is obvious. This approach frames the commons as a blueprint utopia, which in some interpretations shares much with the views of proponents of market environmentalism.

In contrast, the activist perspective tends to frame the commons as a fictional utopia: an inclusive mode of collective stewardship of shared resources, from fisheries and the atmosphere to forests and water. The hope is that forms of commons management, such as public trusts, will encourage a flourishing of the communitarian impulses assumed to be basic to human nature. Markets and states—both depicted as corporate bureaucracies—have little or no role to play. Rather, commons are a set of potentially universally applicable communal social relations, the threats to which have intensified in an era of neoliberal globalization, and the defense of which (via political struggle) is necessary to sustain viable alternatives to capitalism.[5] In some formulations, these alternatives are articulated with a strong sustainability perspective that posits a limit to environmental exploitation (whether phrased in terms of the limits to growth, carrying capacity, or stocks of "natural capital"). This approach is obviously characterized, as I argued, by contradictions: advocates argue that the notion of the commons moves us beyond the stale debate over states versus markets, but the opposition of commons (or communities) to commodities often merely resuscitates the state (public)–market (private) binary.

James McCarthy rightly notes that the disjuncture between these two conceptions of the commons should not be overly troubling. For the broader public, the notion of the commons serves as a fictional rather than a blueprint utopia.[6] Its message is simple: communities can successfully govern resources in a cooperative manner, and we should not assume that resources can successfully be governed only via states or markets. The record of the state in encouraging redistributive models of resource management, progressive social relations, and redistribution is ambivalent, and the record of the unfettered market is unacceptable. To assume otherwise, as Ostrom rightly warns, is misleading.[7] But to abandon the state as a necessary tool for encouraging redistributive models of resource management, progressive social relations, and environmental protection would be equally misleading.

Water: Political and Biopolitical

As I argued in chapter 7, water is simultaneously political and biopolitical. As a flow resource, water transgresses geopolitical boundaries, defies jurisdictions, pits upstream against downstream users, and creates competition between economic sectors both for its use (that is, quantity) and for its disposal (the preservation or degradation of its quality). Water is thus intensely political in a conventional sense: it is implicated in contested relationships of power and authority. But, as explored in chapter 7, water is also biopolitical; it connects individual bodies to the collective body politic through the cycling of waters and water-borne effluents from one watershed and one organism to another, and is the basis for a set of strategies of normalization, state regulation, and self-policed control of health, hygiene, and water-use practices. As I pointed out, the biopolitical nature of water is highly heterogeneous: the attempt to regulate populations becomes constrained by the limits of the developmental state, and by a dual process of inclusion-exclusion that arises as a result. Governments administer populations (Chatterjee's "politics of the governed"); only a small minority are true citizens, treated as rights-bearing individuals. Access to services such as water thus becomes the subject of continuous political negotiation: in most cities around the world, water access is a question of political struggle, often enacted by self-organized community groups operating outside the bounds of traditional pathways of democratic representation, in permanent negotiation with governments over demands for political recognition via the provision of public services.

I thus argued that in seeking equitable solutions for urban water supply we should focus on the creation of conditions for successful mobilization of "political society" (and not the relative performance of public or private providers). Simply put, the challenge is to gain recognition of legal rights and, further, to transform them into social rights (my "material emblems of citizenship"). In much of the world, uneven capitalist development, persistent social cleavages, and unresponsive states remain the key obstacles to this transformation. But there are promising examples in Asia, Africa, and Latin America to which we can turn for inspiration.[8]

This point of view also has implications for our analytical frameworks. If we accept the concept of "political society," then conventional metrics—such as voter participation rates—predicated on concepts such as "civil society" can only partially explain why some communities are successful in gaining recognition and services, and not others. In other words, the search for the reasons behind Porto Alegre's successful management of the urban water-supply problem (and the related question of whether its success can be replicated) cannot focus solely on concepts derived from classical political theory. Other conceptual frameworks are required. In the case of urban water, concepts from political ecology and postcolonial studies may be useful (as in this book) in the development of explanations for the exclusion of the poor from water supply access. Chapter 4, for example, demonstrated how access to water services in Jakarta had its origins in the city's colonial past and its postcolonial politics: urban infrastructure became the subject of political negotiation over urban identities, reflected in the differentiation of urban space and water access. The case of Jakarta suggests that water's dual nature—political and biopolitical—implies the need for government action at higher scales of governance. The question is not *whether* governments should be involved, but how. And, as the book's emphasis on governance sought to underscore, this in turn raises questions of citizenship, identity, and political power to distribute access to social services under the aegis of modern developmental states.

REVISITING THE SOUTH AFRICAN CASE

A specific example may help make the real world implications of these arguments more obvious. We might briefly revisit the example of South Africa, discussed in previous chapters,[9] where governments are struggling to extend water supply services. In urban areas, experiments with private sector participation have had mixed results and have inspired dramatic protests in Nelspruit and Johannesburg. After an initial wave of interest, private water companies have beaten a strategic retreat from new concession contracts, although they remain active with management contracts. Fierce debates rage about how much water, and at what price, governments should provide to the poor. In the meantime, many people are forced to rely on informal private companies or self-provisioning of varying (often poor) quality.

Currently, experimentation is ongoing with a range of community alternatives—from participatory policymaking to citizen participation in regulation to community involvement in sanitation provision. This is particularly evident in rural areas, where communities have a long history of self-provisioning—under apartheid, many communities (usually black and poor) had no choice. In some instances, these rural community systems are relatively equitable and sustainable. But some resist extending these models of community involvement, in part because they involve private, nongovernment actors (and are thus perceived as a type of backdoor privatization). And others argue that citizen participation strategies are often highly manipulative, particularly where they foreclose the pursuit of legal mechanisms to require local governments to fulfill their constitutional obligations.

As discussed in chapter 5, court cases launched to raise the legal minimum for water and sanitation services have, to date, failed. Despite the human right to water in the South African Constitution, the implementation of water-supply and sanitation policies remains relatively constrained. Nonetheless, South African standards are higher than in many other countries, and have risen over the past ten years (for example, the implementation of various Free Basic Water policies). Here, as Patrick Bond and Jackie Dugard argue, legal battles backed by "justice-based rights rhetoric, particularly in the context of struggles against the state, have [had] a beneficial impact in unveiling core contradictions behind commercialization and other cost-recovery related state malfeasance in the [South African] water sector, and enhancing local accountability and responsiveness to community needs."[10]

In urban areas, a range of community solutions has gained popularity, particularly with respect to sanitation. In the absence of a sewerage network, municipal governments (and the private companies to whom they have contracted the service provision) have been encouraging alternatives such as pit latrines, or "condominial shallow sewage" systems. The burden of cleaning the buckets, or the narrow pipes that easily clog, was until recently left to community members (usually women). This situation is not new: bucket latrines are a legacy of the underprovision of services under apartheid. But fifteen years after the end of apartheid (and despite promises that these systems would disappear), nonnetworked sewage technologies remain widespread in urban areas.

In the midsize city of Durban, for example, an estimated tens of thousands of people use bucket latrines, unventilated pit latrines, and so-called urine diversion toilets. The resultant pollution of the city's watercourses is serious and persistent and (particularly in periods of heavy rain and hot temperatures) creates serious public-health risks, especially gastrointestinal diseases that are not just a threat to the healthy but also potentially fatal to the high proportion of the population living with HIV/AIDS, as revealed in recent cholera epidemics. Recent years have seen the extension of some sewerage networks, and of sewage education programs for areas still without service. Here, too, activists argue that community solutions are simply a strategy for unloading government responsibilities onto the poor, while (when private firms are involved) turning a public-health catastrophe into an opportunity for profit. Others argue that private expertise is desperately needed, considering the desperate need for access and limited government resources.

The South African case demonstrates more general points made throughout this book: that the role of communities in water supply is much more ambivalent than many commentators would admit; that human rights are an important tool to be used in fighting for government accountability, but that they cannot, alone, create the conditions for sustainable water access for all; and that issues of ecological and social justice are inseparable, particularly for the urban poor.

PRACTICAL IMPLICATIONS

With this in mind, I now turn to a discussion of practical implications of my arguments for the water privatization debate. The first point is that we need to look beyond the focus of current debate on the need for financing (the "financing gap"). This is not only because the total amounts of financing required are, in global terms, relatively modest. The cost of financing new household connections for water and sewerage to meet the modest target of 75 percent coverage of urban areas in developing countries is estimated to be US$80 billion—one-third of the estimated cost to date (mid-2009) of the Iraq War, and approximately 5 percent of the total guarantees provided to banks by G7 countries.[11] It is also because the availability of financing will not address the governance challenges we have identified.

Second, to be effective, the human right to drinking water will need to be associated with reforms to community water rights: the human right to water and water rights must be considered together, both to avoid potential conflicts (as in the case of indigenous water rights) and to better foster the expanded notion of rights advocated in this book. At the same time, "rights talk" needs to be articulated with—and indeed constrained by—ecological rights: the concept of ecological governance, as I argued in earlier chapters, might allow us to do this. With this in mind, fulfilling these rights will require a range of technical measures, including, for example, effective monitoring and data collection (from which informal settlements are often excluded) on water quality and access, not only to avoid water-borne diseases but also in order to hold governments and other water providers accountable for their obligations.

Third, the role of private actors needs to be rethought, as the Porto Alegre example at the start of this chapter was intended to suggest. Private sector provision, particularly in the forms advanced in the 1990s, cannot sustainably supply water to the urban poor, much less address broader issues of water governance; nor can private capital finance network extensions on a significant scale. But private actors do have a role to play (in providing, for example, expertise and innovative technologies). All of this suggests that, as in the Porto Alegre case, our approach to the division of responsibility among governments, private actors, and communities, and the appropriate goals and governance mechanisms that should guide them (some of which may be market oriented) should be pragmatic. This does not imply, as some might think, an unprincipled strategy. Earlier chapters suggested some of the principles that are likely to guide communities: inclusive participation, transparency, accountability, and environmental conservation. In making these choices, communities must have a voice, the playing field must be level, and privatization must not be coerced or forced on communities.

This implies the need for action at multiple scales: from negotiations over international trade treaties (such as the General Agreement on Trade in Services and bilateral investment treaties to open water markets to foreign investment in competition) to regional and local governments. A broader approach at the international level (such as a Framework Convention on water under the UN system) might be desirable. Here, national governments will be the key actors: their scale and legitimacy enable them to intervene in the global legal and economic frameworks that have played such an important role in shaping local conditions around water privatization and water sector reforms more generally.

This reasoning anticipates my fourth point: we should be realistic, and cautious, about community water management in its various incarnations. It is true that community water-management models have a greater potential to be appropriately inclusive and environmentally sustainable, as well as sensitive to preexisting traditional water rights. But it is equally true (as explored in chapter 6) that there is a range of issues that local communities cannot resolve, particularly in urban areas: the resolution of upstream and downstream conflicts within large watersheds, for example. Recourse to the state (despite unresponsiveness, corruption, and inefficiency that may impair public utilities) is necessary and can be productive where communities are able to create the conditions for successful struggles for access and control of water supply as well as for sustainable environmental living conditions.

Such is particularly the case with respect to small-scale independent providers, which have been the focus of increased interest following the retreat of the international private sector from the poorest regions and cities of the world. The roles played by these providers were explored in chapter 4. Support for these "alternative private providers" comes from those broadly who favor private sector involvement (as a continuation of privatization by other means) and from some in favor of community-based development solutions, often in tandem with "appropriate" (i.e., small-scale and locally adapted) technologies. Accepting the interim existence of these small-scale providers is realistic. But supporting their entrenchment in water supply governance is, I suggest, a mistaken endorsement of an unsatisfactory status quo that misguidedly overlooked the necessary role of public policy in correcting market imperfections. Poor service quality, the loss of economies of scale, and the entrenchment of dual or multiple water-supply delivery standards associated with socioeconomic status are serious concerns.[12] The most equitable long-term solution is universal provision, overseen by the state.

These measures will not forestall conflict over water privatization, or indeed water more generally. Water is essential for life, and so imbued with symbolic meanings: purity, divinity, and health. It supports the ecologies on which we depend. As a resource, it is also an essential input for our economies: our industries, agriculture, and cities are lubricated with water. Waterways are sources for drinking water yet conduits for waste. So it is unsurprising that conflicts over the use, ownership, and conservation of water are long-standing.

These issues make the task of water governance a difficult one. Societies confront the simultaneous need to mobilize water resources for a range of uses, to mediate conflicts between users and uses, to build and maintain infrastructure, and to dispose of wastes, while mediating jurisdictional scales. Water inevitably inspires conflict, of which privatization is simply one example.

So while we will never transcend these conflicts, we can still hope to resolve them in more equitable and ecologically sensitive ways. This will inevitably be a contested process, in which claims are mobilized against both governments and private capital. Here, perhaps, is where my analysis is in agreement with the utopian views that have been the subject of much critique in earlier chapters. Where I differ is in my views on appropriate alternatives. The failure of technocratic high modernism, which characterized public water supply in developing countries throughout much of the twentieth century, is clear. However, neither privatization nor atomized community solutions are viable responses. The choice with which the public-private debate presents us (between neoliberalism and anarcho-communitarianism) is false. Rather, our analyses (and political struggles) are better directed toward the terrain of "political society": the search for models that resolve, to the extent possible, the inevitable tensions between representation and participation, technocracy and democracy, centralized oversight and local preferences, and economic exigencies and environmental imperatives. The key actors in these conflicts will be communities and the governments they attempt to hold to account. This does not imply the total exclusion of private actors from water supply provision. On the contrary, it requires their enrollment, but—as I have sought to suggest—on very different terms than in the past.

Notes

Introduction

1. Conca 2006.

2. See the World Bank's 1993 *Water Resources Management Policy Paper* for an early and influential formulation.

3. Lobina and Hall 2008.

4. Harvey 2003a.

5. Barkin forthcoming; Petrella 2001; Shiva 2002.

6. Harvey (2005) characterizes neoliberalism as a set of ideas that has dominated public-policy debates since the 1970s, at the core of which is the doctrine that market exchange is an ethic thought to be capable of acting as a guide for all human action.

7. See Balanyá et al. 2005; Birdsall and Nellis 2003; Castro 2008; Davis 2005; Hall, Lobina, and de la Motte 2005; Prasad 2006; Shiva 2002; WDM 2006.

8. In this chapter, since many readers will be unfamiliar with the debate, I use the term "privatization" as a shorthand for a broad range of private sector involvement. This is technically inaccurate. See discussion of the term "privatization" on pp. xv–xvi.

9. Because of the lack of reliable data, current estimates of urban dwellers without access to adequate provision for water supplies are necessarily highly uncertain, but it seems likely that the true numbers of urban dwellers with inadequate provision for water supply are significantly underestimated. According to the UN Centre for Human Settlements, the cumulative total of urban dwellers without access to adequate water supply is 970 million (for Africa, Asia, Latin America, and the Caribbean) (Gleick 1998; UNCHS 2003, 2006; WHO 2003). The World Health Organization and UNICEF Joint Monitoring Programme for Water Supply and Sanitation provide figures for the proportion of the population with "improved" provision, including water from public standpipes, boreholes, and protected dug wells, provided that at least twenty liters per person per day is available from a source within one kilometer of the person's home. This definition does not account for whether the water is "adequate" or "safe" to drink. The question of the level of "adequate" water supply is, however, the subject of

dispute. Although the World Health Organization recommends one hundred liters per person per day, other experts have proposed fifty liters per person per day as a minimum standard to meet four basic needs—drinking water, sanitation, bathing, and cooking. (As a comparison, North Americans use closer to four hundred liters per person per day.) (Gleick 1996; UNCHS 2003, 2006; WHO 2003a).

10. Many other important issues are at play: water for irrigation, for example, represents 70 percent of fresh-water use in many places. And many rural dwellers lack access to adequate amounts of safe water.

11. See "Defining 'Privatization,'" pp. xv–xvi.

12. Winpenny 2003.

13. As of January 2006, 11 percent of the world's large cities (population greater than 1 million) had private sector participation contracts in place (Lobina and Hall 2008). As with many other aspects of the debate over privatization, there is fierce disagreement over the precise measure of customers of private sector companies. Perhaps 10 percent of cities with populations larger than three hundred thousand in lower- and middle-income countries have private sector participation (2006 data). These tend to be larger cities, representing approximately 20 percent of the world's urban population. See Hall and Lobina 2006b. The figure of 3 percent is cited in Winpenny 2003.

14. See footnote 6 above.

15. Anderson and Leal 2001; Bernstein 2001; World Resources Institute 2008.

16. Heynen et al. 2007; Himley 2008; Mansfield 2008.

17. O'Connor 1996.

18. This is true of the literature on public services more generally: many well-documented cases exist where public authorities do not serve the needs of poor families, but case studies also exist of well-performing public services (see Castro 2007; Tendler 1997). With specific reference to water supply, there is little agreement on the bases on which studies of the performance and impacts of government versus the private sector can be conducted. Some of the disagreements are methodological; for example, the choice of cross-sectional econometric analysis versus detailed case studies. In addition, studies have been plagued by a lack of data (particularly with respect to reliable baseline information to enable the evolution of parameters). The choice of parameters is also important (efficiency, connections, tariffs, etc.); opponents and proponents of PSP have tended to select different parameters, thereby producing competing (and often divergent) evaluations of the same cases.

19. Taylor 2004.

20. Chatterjee 2004.

21. See Castro, Kaika, and Swyngedouw 2002; Harvey 1996; Keil and Desfor 2004; Swyngedouw 1997, 2004.

22. Water supply was a central focus of the key environmental meetings in the 1970s, including the Stockholm Environment (1972) and Vancouver Habitat

(1976) conferences, and the UN's Mar del Plata conference (1977), at which the Water and Sanitation Decade was formulated (WHO 1992).

23. During the decade, it was estimated that 1.2 billion people gained access, for the first time, to a safe and adequate water supply, but the number without access to a safe and adequate water supply fell by only 450 million during the same period (WHO 1992). However, statistics on access to water supply should be treated as estimates. From approximately 1960 to 2000, data were gathered through sector monitoring based on national reporting overseen by the World Health Organization, later joined by UNICEF. Since 2000, data have been gathered through nationally representative household surveys ("coverage monitoring"). The discrepancies in coverage statistics between the two approaches are high, given that one is a "supply" and the other is a "user" perspective. (J. Bartram, WHO, personal communication).

24. Grusky 2001; Hall and de la Motte 2004.

25. According to the UN Population Fund, in 2005, 41 percent of the population of less- or least-developed countries (2.7 billion people) are considered "urban," and 50 percent of the world's total population (including developing countries) are now urban. As precise measures of population are difficult, we should recognize the approximate nature of these figures.

26. This is not to diminish the importance and urgency of rural water supply issues. The focus on urban water is merited by the fact that differences between urban and rural water supply are significant, and the focus of formal water privatization on urban areas is so overwhelming.

1. Governance Failure

1. Graham 2002; Graham and Marvin 2001.

2. Brennan and Richardson 1989; Chatterjee 2004; Leitmann 1995; McGranahan et al. 2001; McIntosh 2003.

3. Alsayyad and Roy 2006; Roy 2005.

4. See Appadurai 2002; Chatterjee 2004; Davis 2006.

5. I am indebted to Roger Keil for these insights.

6. Weintraub and Kumar 1997.

7. Foucault 1991.

8. Chatterjee 2004; Stoler 1995.

9. Chatterjee 2004.

10. Mansfield 2008.

11. Bakker 2004.

12. Bakker 2004; Coutard 1999; del Moral and Sauri 2001; Goubert 1989; Hassan 1998.

13. A few important exceptions are frequently mentioned in the literature, such as private (but highly regulated) companies in the water supply industry

in France and England and water markets in the United States, Chile, and the Canary Islands.

14. The history of private sector involvement in water supply provision in the nineteenth century is discussed in chapter 3.

15. Biswas 2004; Gleick 2000; Graf 2001; Ortolano and Cushing 2002; WCD 2000.

16. See, for example, the U.S. Clean Water Act (33 U.S.C. 1251–1376) and the EU Water Framework Directive (Directive 2000/60/EC).

17. Kinnersley 1988; Melosi 2000a, 2000b.

18. Gleick 2000; O'Connor 1973.

19. Another justification in some countries was the fact that water use began to decline. This was in part due to deindustrialization, but also to increasing technical efficiency in an era of heightened concern over resource scarcity and changing patterns of domestic water use. In some countries, such as the United Kingdom and the United States, water withdrawals began declining in the mid-1980s and by 2000 were 10 percent below their peak. See Gleick 2000.

20. Kaika and Swyngedouw 2000.

21. Robertson 2000, 2004.

22. Graf 2001; Holloway 1994.

23. Bakker 2003a, 2003b; Graf 2001; Huffaker and Whittlesey 2003; Kijne 2001; Kloezen 1998; Kumar and Singh 2001; Landry 1998; Takahashi 2001; Ward and Michelsen 2002.

24. Cesano and Gustafsson 2000; Gleick 2000; Rogers, Silva, and Bhatia 2002; Saleth and Dinar 1999; Winpenny 1994.

25. Lacey 2004.

26. Bakker 2004.

27. Bakker 2008; Boelens 2009; Boelens and Gelles 2005; Ratner 2004.

28. Ostrom 1990; White, Bradley, and White 1972.

29. Mehta 2005.

30. Boelens and Gelles 2005.

31. Boelens and de Vos 2006; Perreault 2008.

32. Scott 1998.

33. Indeed, I would argue that Scott falls into this trap himself in his seminal work on the subject, *Seeing Like a State* (Scott 1998). See also Ellul, 1964.

34. Li 2007.

35. Boelens 2009.

36. They are, in contrast, richly documented in the academic literature. See Boelens 2008.

37. For an exception, see Kjellén 2006.

38. Market failures include ill-defined or unenforceable property rights, "public goods" (when goods are nonexcludable and nonrivalrous), externalities (when prices do not incorporate full costs or benefits), information asymmetries, and monopoly.

39. Jessop 2006, 381.

40. Pierre 2000; Rhodes 1996.

41. Adapted from Rogers and Hall 2003 and UNDP 2007.

42. Pierre and Peters 2000.

43. Jessop 1995; Pierre 1995, 2000; Pierre and Peters 2000; Rhodes 1996.

44. Swyngedouw (1997) terms this a process of "globalization."

45. See Faust and Nagar 2003.

46. McCay and Jentoft (1998) refer to something similar when they discuss "community failure."

47. Jessop 2006, 381.

48. Ahmed 2008.

49. Appadurai 2002.

50. Budds and McGranahan 2003.

51. In Sen's terminology, "capability" is an individual's ability to achieve desired "functions," where functions represent actual achievements in "doing or being"—both tangible and intangible (for example, being well nourished, adequately hydrated, healthy, well educated, or having self-respect due to active participation in community decision making). A "capability set" is defined as the set of attainable functions that an individual is able to achieve. The concept of a capability set is analytically useful because it provides a means of explaining why different individuals are able to differentially mobilize specific commodities to achieve certain functions. See Sen 1992, 1999.

52. Ibrahim 2006.

53. Chatterjee 2004, 37–38.

54. Harvey 2003b; Mitchell 2003.

55. Chatterjee 2004, 40.

56. The "world cities" debate is the lens through which the question of urban primacy has often been framed in recent years. For a critique of this work and a discussion of alternative strategies for studying "ordinary cities," see Amin and Graham 1997; Robinson 2006. For an earlier discussion, see Gilbert and Gugler 1992.

57. See UNDP 2007.

2. Material Emblems of Citizenship

1. Goubert 1989.

2. Graham and Marvin have termed this the "modern integrated ideal" (Graham and Marvin 2001) of urban infrastructure provision.

3. See, for example, Boelens 2008; Roth, Boelens, and Zwarteveen 2005.

4. See, for example, Boelens 2008.

5. Goubert 1989; Illich 1985.

6. D'Souza 2006.

7. Lubis 2004.

8. Goubert 1989, 229.

9. Gandy 2002, 1999; Goubert 1989; UNDP 2006.

10. Tarr 1996.

11. Cipolla 1992; Goubert 1989.

12. Tarr 1996; Melosi 2000a, 2000b.

13. See Kaika 2005 for a discussion of the "fetishization" of urban water supply networks.

14. Scudder 2005.

15. Gleick 2000, data table 16.

16. Gerschenkron 1962; Habbakuk 1961; Kuznets 1966; Rostow 1960.

17. Meier 2005.

18. Ward 1997, 47.

19. Van De Giesen et al. 2001; Hughes 2006.

20. Swyngedouw 2004.

21. Shiklomanov 1997.

22. "Green revolution": The transfer of technology (largely coordinated through national agricultural agencies) that sought to increase production through an industrialization of agriculture, employing purpose-bred seed varieties designed to produce high yields in association with intensive fertilizer and water inputs.

23. Postel 1999.

24. WCD 2000.

25. Ibid.

26. Goldsmith and Hilyard 1984; Khagram 2004; McCully 1996; Von Tunzelmann 1978; WCD 2000.

27. Birdsall and Nellis 2003; Biswas 2004; Graf 2001; Ortolano and Cushing 2002; WCD 2000.

28. Dubash et al. 2001; WCD 2000.

29. WCD 2000, xxviii.

30. McCully 1996; Rich 1994.

31. WCD 2000.

32. Grossman 2002.

33. Created in 1948. See Staples 2006.

34. Kapur, Lewis, and Webb 1997.

35. Mason and Asher 1973.

36. Kappagoda 1996; Peprah 1994; Tussie 1995. For example, IDB lending for water supply to Latin American borrowers from 1960 to 1970 was 2.5 times the combined amount lent by the World Bank Group (IBRD and IDA) for water supply globally over the same period (Mason and Asher 1973).

37. These included the UN Economic Commission for Africa; the Inter-American Development Bank; the UN Special Fund for Economic Development (later UNDP); the Development Assistance Committee of the Organization for Economic Cooperation and Development; as well as several bilateral aid agencies (Kapur, Lewis, and Webb 1997, 151–152; World Bank 1992b).

38. Mason and Asher 1973.

39. A parastatal is an organization, agency, or corporation partially or wholly owned by government, but operating independently of direct state control.

40. Coutard 1999; Moral and Sauri 2001; Kaika 2005; Worster 1985.

41. Mason and Asher 1973, 190.

42. Blokland, Braadbaart, and Schwartz 1999.

43. UNWWAP 2003, 2006.

44. See World Bank 1993 for an influential discussion. See also Spiller and Savedoff 1999.

45. World Bank 1994.

46. UNDP 2006.

47. Mason and Asher 1973.

48. Kapur, Lewis, and Webb 1997, 393.

49. Mason and Asher 1973.

50. Kapur, Lewis, and Webb 1997.

51. Rich 1994, 318.

52. Staples 2006.

53. Sadove and Squire 1960.

54. World Bank internal memo, P. Squire to R. Sadove, September 23, 1960. This and other internal memos of the World Bank cited in this book are housed in the World Bank archives, and were consulted in June 2000.

55. World Bank internal memo, M. Hoffman to A. Kamarck, March 21, 1966.

56. This logic is comparable to the memo from World Bank chief economist Lawrence Summers in 1991, which argued that exporting polluting industries and waste products from rich to poor nations made sense because low-income countries were "under-polluted." For a more detailed discussion of the Summers memo, see Hausman and McPherson 2006.

57. Mason and Asher 1973, 310.

58. WB and IDA 1971. See also Kapur, Lewis, and Webb 1997.

59. Mason and Asher 1973.

60. Ibid., 152.

61. Rich 1994, 190.

62. Mason and Asher 1973, 227.

63. Kapur, Lewis, and Webb 1997, table 7–3 and table 4–2.

64. Ibid., 201.

65. Ibid., 258.

66. IBRD 1971.

67. Mason and Asher 1973.

68. Kapur, Lewis, and Webb 1997.

69. Mason and Asher 1973, 258.

70. Black 1998, 298; Pitman 2002.

71. World Bank 1989a.

72. The loan was approved on December 15, 1977 and was the first loan entirely for rural water supply and waste disposal (loan 1502 Rural Water Supply Project in Paraguay). Online archives of the WB at http://web.worldbank.org/wbsite/external/extaboutus/extarchives/0,cntrymdk:82511~menupk:64319211~pagepk:64319278~pipk:64321441~thesitepk:29506,00.html.

73. Evans and Appleton 1993; Kapur, Lewis, and Webb 1997.

74. Rich 1994. See also http://www2.irc.nl/manage/whatisit/history.html.

75. Morse and Berger 1992; Wood 1993; World Bank 1992a. The Morse team's independent review was commissioned by former World Bank president Barber Conable and the Bank's board of directors after years of controversy over resettlement and environmental impacts of the project. The Bank withdrew from the Sardar Sarovar (Narmada) project in 1994.

76. World Bank 1992a.

77. The so-called Wapenhans Report (officially titled "Effective Implementation: Key to Development Impact; Report of the Portfolio Management Task Force") argued that pressure to lend resulted in excessive attention to loan making and insufficient attention to implementation and supervision of loans, resulting in a higher-than-necessary rate of project failure—approximately 20 percent at the time of the report (see World Bank 1992b).

78. World Bank 1992b, 43.

79. World Bank 1992b.

80. World Bank 1992a, table 1.2.

81. World Bank 1992a.

82. Ibid., box 5.1.

83. World Bank, 1992a, section 5.15.

84. EDF 1992.

85. Kapur, Lewis, and Webb 1997.

86. World Bank 1993.

87. Kapur, Lewis, and Webb 1997, 31.

88. The 1992 International Conference on Water and the Environment set out what became known as the "Dublin Principles," including principle no. 4 that "water has an economic value in all its competing uses and should be recognized as an economic good." The three other principles were (1) "Fresh water is a finite and vulnerable resource, essential to sustain life, development and the environment"; (2) "Water development and management should be based on a participatory approach, involving users, planners and policy-makers at all levels"; and (3) "Women play a central part in the provision, management and safeguarding of water." The Dublin Principles have been adopted by numerous international, multilateral, and bilateral agencies.

89. Miller-Adams 1999.

90. Megginson and Netter 2001, 323. See also Nellis 2002.

91. Miller-Adams 1999.

92. Ibid.

93. However, on-lending to the private sector via other organizations (such as national banks) was a long-standing practice.

94. The World Bank Group is the term used to refer to the World Bank and its affiliates, the International Finance Corporation and the International Bank for Reconstruction and Development.

95. World Bank 1989b.

96. Miller-Adams 1999.

97. World Bank 2002.

98. Chang and Rowthorn 1995; Bangura and Larbi 2006.

99. Williamson 2000.

100. Saad-Filho and Johnston 2005.

101. Stiglitz 1998.

102. Head 1996; Weber 1994.

103. Official development assistance (ODA) includes grants and loans to countries (directly or via multilateral institutions) that are administered in order to promote economic development. For a precise definition, see the OECD Development Assistance Committee website, which also provided the ODA data referred to in this section: www.oecd.org/stats/idsonline.

104. Total ODA commitments declined from 1991 to 1997, after which they began rising slowly, accelerating only after 2002, spurred by the new commitments made at the Monterrey summit in March 2002. However, some of this rise has been due to debt rescheduling and reduction, and the actual growth rate may be overestimated by the decline in the U.S. dollar (in which ODA is measured) relative to the yen and euro—as Japan and the euro zone together provide approximately 60 percent of ODA. This was influenced by a number of factors—notably, the perception on the part of some donor countries of the declining geostrategic importance of aid in the post–cold war context, combined with fiscal discipline that sometimes reduced the scope for aid flows. Some observers have argued that another contributing factor, in the case of water, was the expectation that increased private foreign direct investment would compensate.

105. WWC 2005.

106. World Bank 2000.

107. For example, Stiglitz 1994.

108. Batley and Larbi 2006; Stiglitz 2003.

109. Miller-Adams 1999.

110. Most notably Britain's Department for International Development and the Canadian International Development Agency.

111. World Bank 2003a.

112. Swyngedouw 2004.

113. See Chatterjee 2004. This also evokes the "social capital" debates, on which see Evans 1995, 1996; Fine 2001; Putnam 2000; Tendler 1997; Wade 1990.

3. WATERING THE THIRSTY POOR

1. According to the subsequent legal ruling by the International Center for the Settlement of Investment Disputes, the government of Tanzania "was awarded World Bank, African Development Bank and European Investment Bank funding in the amount of USD 140,000,000 for the purpose of commissioning a comprehensive program of repairs and upgrades to, and the expansion of, the Dar es Salaam Water and Sewerage Infrastructure. As a condition of the funding, the Republic was obliged to appoint a private operator to manage and operate the water and sewerage system, and carry out some of the works associated with the Project." (ICSID 2008, 1).

2. Hall and Lobina 2006b.

3. See, for example, ActionAid 2004.

4. Biwater 2005; Rice 2007.

5. ICSID 2008.

6. Biwater invoked its rights as a foreign investor under a UK-Tanzania bilateral investment treaty (signed in 1994) to initiate the arbitration against Tanzania.

7. ICSID 2008.

8. The case was brought by the Dar es Salaam Water and Sewerage Authority against City Water Services, and was heard in London under the rules of the UN Commission on International Trade Law (UNCITRAL).

9. See Kjellén 2006.

10. Vidal 2005.

11. See, for example, McDonald and Ruiters 2005.

12. See, for example, Franceys 2008.

13. More rarely, greater equity (through directing water to its "highest-value uses") and greater environmental protection (largely through water conservation associated with full-cost pricing) are invoked.

14. See "Defining 'Privatization'" in the introduction; also see OECD 2000 for an alternative, and arguably more pro–private sector historical survey.

15. Melosi 2000a, 2000b.

16. In France, private companies managed approximately 70 percent of the country's municipal water-supply systems on a contractual basis (ownership was retained by municipalities). In Britain, small private companies were permitted to continue owning and operating water supply infrastructure, although the majority of consumers were supplied by publicly owned and managed water authorities. These two models were to form the basis of the twentieth-century experiment with private sector provision.

17. In Britain, for example, the number of municipal water undertakings increased from eleven in the 1830s to over seven hundred in 1914, by which time local governments controlled over 80 percent of water supply systems (Millward

1989; Sheail 1983, 1986). A similar pattern developed in the United States (Melosi 2000a, 2000b).

18. Cutler and Miller 2006.

19. Dracup 1973.

20. Bakker 2004.

21. Goubert 1989.

22. Bakker 2004; Coutard 1999; Goubert 1989; Hassan 1998; Melosi 2000a, 2000b.

23. Public financing was critical for the development of water supply in most countries. In the United States, for example, capital investment for water over the twentieth century is estimated at $400 billion (unnormalized) (Rogers 2003). During the New Deal period alone, the Public Works Administration financed over twenty-five hundred water projects (Melosi 2000b). Other industrialized countries show similar patterns. In Spain, for example, a country with one of the highest proportions of surface area covered by reservoirs in the world, the nominal economic value of the water resources made available through hydraulic development has been calculated at a range of between 5 percent and 8 percent of GDP (Martín Mendiluce 1996; MIMAM 1998).

24. Chant 1989; Graham and Marvin 2001.

25. Cutler and Miller 2006.

26. Gregg 1950.

27. Millward 1989; Sheail 1983, 1986.

28. Cutler and Miller 2006.

29. Privatization occurred through flotation on the London Stock Exchange. See Bakker 2004 for a detailed analysis of the British case.

30. UNWWAP 2003, 2006.

31. For a discussion of the standard range of private sector participation contracts, see Budds and McGranahan (2003). These reforms refer solely to water supply management. Reforms also occurred with respect to water resources management; for example, the creation of water rights and water markets.

32. Banerjee, Oetzel, and Ranganathan 2006; Jensen and Blanc-Brude 2006.

33. Williamson 2000.

34. Bangura and Larbi 2006; Stiglitz 2003. For a detailed critique of the debt problem, see George 1988, 1992.

35. Rich 1994.

36. Bernstein 2001.

37. For the text of the Dublin Principles, see chapter 2, note 88.

38. See, for example, Goldman 2005; Robertson 2004, 2007; Shiva 2002.

39. For a summary of critiques of the World Bank's PPI database, see Hall and Lobina 2006a.

40. Marin and Izaguirre 2006.

41. Izaguirre 2005.

42. Whereas 9 percent of the total investment flows for infrastructure projects with private sector involvement were canceled between 1990 and 2004, the equivalent figure for the water sector was 37 percent (Annez 2006; Izaguirre 2005). This figure is, however, skewed by the cancellations of several large projects. Calculated on the basis of numbers of projects, the cancellation rate from 1984 to 2007 was 9 percent (Marin 2009).

43. Whereas telecommunications attracted 70 percent of investment (peaking in 1998), the water sector accounted for only 3 percent of investment (Izaguirre 2005).

44. OECD 2000.

45. Winpenny 2003; World Bank/IMF 2003.

46. The most recent study by the World Bank suggests that the global population served by private companies for urban water supply increased from approximately 96 million in 2000 to more than 160 million by 2007 (Marin 2009).

47. Palaniappan et al. 2006.

48. Marin 2009.

49. Kerf and Izaguirre 2007.

50. WDM 2005.

51. WHO 2008.

52. Marin and Izaguirre 2006; WDM 2005.

53. Davis 2005; OECD 2009; Prasad 2006.

54. Lobina and Hall 2008; Mehrotra and Delamonica 2005.

55. Estache, Guasch, and Trujillo 2003; World Bank 2003a.

56. Banerjee, Oetzel, and Ranganathan 2006; Franceys and Gerlach 2008; Groom, Halpern, and Ehrhardt 2006; Jensen and Blanc-Brude 2006.

57. Bakker 2007a; U.S. GAO 2002; NRC 2002; World Bank 2006b.

58. Hall, Lobina, and de la Motte 2005.

59. Bennett, Dávila-Poblete, and Rico 2005; Morgan 2005; Olivera and Lewis 2004.

60. World Bank 2005, 22.

61. Izaguirre 2005, note 299.

62. PricewaterhouseCoopers 2003.

63. Nellis 1999; Sachs 1999.

64. Cour de cassation—chambre criminelle du 08/04/1999, pourvoi n\060 98–84539, Jean-Jacques Prompsy et autres; Tribunal Administratif de Grenoble du 12 mai 1999, n° 982087, M. Comparat; Tribunal Administratif de Grenoble du 07/08/1998 no 962133, 964778, 964779, 964780, 98481 et 98482, M. Betto et autres. Cour de Cassation 1999.

65. Franceys 2008.

66. Estache, Gomez-Lobo, and Leipziger 2001.

67. I am grateful to Patrick Bond and the Centre on Housing Rights and Evictions for pointing this out.

68. Thompson 1971.

69. Davis 2005; Prasad 2006.

70. Marin 2009.

71. Balance and Trémolet 2005.

72. Clarke, Kosec, and Wallsten 2004.

73. See, for example, WDM 2006.

74. World Bank 2005; IMF 2004, 3.

75. World Bank 1997a.

76. For example, Estache and Rossi 2002; Megginson and Netter 2001.

77. Bayliss 2002; Estache and Rossi 2002; Hodge 2000; Hunt and Lynk 1995; Kirkpatrick, Parker, and Zhang 2006; Lobina and Hall 2000; Prasad 2006; Renzetti and Dupont 2004; Wallsten and Kosec 2005.

78. Holmes 2000; Verhoest et al. 2004.

79. Lobina and Hall 2008.

80. Gassner, Popov, and Pushak 2009, 49.

81. IMF 2004.

82. ADB 2003a; Budds and McGranahan 2003; Franceys and Gerlach 2008; Gutierrez 2002; Robbins 2003; UNDP 2003.

83. Cross and Morel 2005; Galiani, Gertier, and Schargrodsky 2002; Johnstone and Wood 2001; Nickson and Franceys 2003; Shirley 2002; Winpenny 1994; World Bank 1994, 1997a, 2004.

84. Bayliss 2002; Bayliss and Fine 2007; Bond 2004; Bond 2002; Hukka and Katko 2003; McDonald and Ruiters 2005; Wilder and Lankao 2006.

85. Moreover, the difficulties of comparing firm performance are compounded by the difficulty of comparing accounting information across different countries, at different points in time, while controlling for changes in the macroeconomic environment. And short-term negative impacts (such as those on labor) may be mitigated for some firms, and for some employees, over the long term (Gupta et al. 2001).

86. IMF 2004.

87. On the Dutch model, see Blokland, Braadbaart, and Schwartz 1999.

88. In technical language, supporters of public water argue in favour of subsidies for CAPEX (capital expenditure) as well as OPEX (operating expenditure).

89. For example, Collignon 1998; Kariuki and Schwartz 2005; Solo 1999.

90. Weitz and Franceys 2002. See also the Business Partners for Development initiative, BPD 2002, at http://www.bpdwaterandsanitation.org/english/docs/innovation.pdf.

91. Altenburg and von Drachenfels 2006.

92. Baietti, Kingdom, and van Ginneken 2006; Balanyá et al. 2005; Bayliss and Fine 2007; Lobina and Hall 2000; Wolff and Hallstein 2005.

93. Baietti, Kingdom, and van Ginneken 2006; Wolff and Hallstein 2005; Van Hofwegen 2006. UNSGAB has led to the Water Operator Partnerships initiative, which is being taken forward under the auspices of UN-Habitat and the International Water Association, with participation from a broad range of

stakeholders, including major private water operators and public sector unions. The involvement of private companies in the UNSGAB initiative has been the source of much criticism. On public-public partnerships see Balanyá et al. 2005. See also Hall, Lobina, and de la Motte 2005.

94. For example, Ringskog, Hammond, and Locussol 2006.

95. The World Bank Group's Multilateral Investment Guarantee Agency did take the novel step in 2001 of granting "political risk" coverage for a Dutch firm participating in a thirty-year concession contract in Guayaquil, Ecuador. The call in the controversial Camdessus report for official development assistance to be used to "facilitate" private sector activity was never widely applied (Winpenny 2003). This would have implied, for example, the redefinition of ODA. Other strategies included the provision of bilateral grants for network extensions in water supply systems managed by private sector concessionaires through "output-based aid" initiatives (a discussion of which occurs in chapter 5).

96. Arendt 1958; Chatterjee 2004; Habermas 1989; Putnam 2000.

97. Habermas 1989.

98. Weintraub and Kumar 1997.

99. Mansfield 2008.

100. See, for example, Estache 2005; Kessides 2005; World Bank/IMF 2003; World Bank 1994, 2006a, 2006b.

101. Kessides 2005; World Bank 2004.

102. This overlaps academic debates over the role of the state, framed in part through debates on the "new theory of the firm," the role of institutions (the "new institutional economics"), and the role of the "developmental state" in fostering economic growth. See Burki and Perry 1998; Chang 1999; Stiglitz 1998; Wade 1990; Woo-Cummings 1999.

103. Gore 2000; Mehrotra and Delamonica 2005.

4. CITIZENS WITHOUT A CITY

This chapter draws heavily on research conducted during the course of M. Kooy's doctoral thesis, "Relations of Power, Networks of Water: Governing Urban Waters, Spaces and Subjects in (Post)colonial Jakarta," in the Department of Geography at the University of British Columbia (2002-2008). The material presented in this chapter appeared, in modified form, in several previously published articles; permission to reproduce this material is gratefully acknowledged.

1. Supporting a huge growth in the number of small-scale businesses selling generic bottled water, these refill stations provide treated, potable water at less than one-third the price of the brand names. Generic bottled water from the *air isi ulang* (refill stations) is Rp 3,000/19 L, whereas the Aqua brand preferred by the middle class sells for approximately Rp 10,000/19 L. FORKAMI, personal communication, January 10, 2007; Weimer 2006.

2. See Xenos (1989) for a discussion of scarcity under modernity.

3. Ferguson 1994.

4. Li 2007. Ferguson's (1994) evocative phrase for this is the "anti-politics machine."

5. Mitchell 2002.

6. The first figure is from the annual SUSENAS socioeconomic survey conducted by the Indonesia Bureau of Statistics (BPS 2005). The second figure was calculated using data from the regulatory authority overseeing the management of Jakarta's water supply system (Jakarta Water Supply Regulatory Body 2004). This was cross-referenced with an ADB report (2003b) that records 51.2 percent. In Jakarta, coverage ratios are always imprecise estimates; their calculation is dependent on a number of variables that are only imprecisely measured, such as urban population and average size of household. Reported figures vary significantly and do not indicate the number of households that have a connection but rely primarily on other sources (like groundwater) owing to quality or service concerns (such as low pressure). Large numbers of seasonal migrants and informal (officially "illegal," without residence permits) residents without land tenure mean that population figures are systematically underestimated and that, as a consequence, coverage figures are systematically inflated.

7. BPS 2005; Crane 1994; Jakarta Water Supply Regulatory Body 2004; McGranahan, et al. 2001; Shofiani 2003; Woodcock 2006.

8. Tutuko 2001.

9. ADB 2003b.

10. Berry 1982; Gilbert and James 1994; Lovei and Whittington 1993; McGranahan et al. 2001.

11. See, for example, Cairncross 1990; Gulyani, Talukdar, and Kariuki 2005; Swyngedouw 1997.

12. Argo 1999; Cowherd 2002; World Bank 2003b.

13. See, for example, Crane and Daniere 1997; McIntosh 2003; Porter 1996; Surjadi 2003.

14. Braadbaart and Braadbaart 1997.

15. Harsanto 2005; Simamora 2007.

16. Agtini et al. 2005; Leitmann 1995; McGranahan et al. 2001; Simanjuntak et al. 2001; Surjadi 2003.

17. Agtini et al. 2005.

18. This reconstruction of the growth of the city's network relies on primary archival data gathered in 2003 at the Dutch colonial archives in the Netherlands. See Bakker et al. 2008.

19. Van Breen 1916; Van Leeuwen 1917.

20. Eggink 1930; Maronier 1929. Specifically, in 1929, 6,926 kampung households (estimated population 400,000) were supplied with twenty-four liters, while 10,392 European households, a population of 37,067, were supplied with eighty-four liters (Eggink 1930, 930; Heetjans 1923; Maronier 1929).

21. Moens Bernelot 1873.

22. Stoler 1992.

23. Cote 2003; Stoler 1992.

24. The "degenerate" European residents were largely of mixed ancestry; legally considered "European," these residents were the offspring of the long-time colonial settlers who had adopted an Asianized urban culture. See Taylor 1983 and Milone 1967.

25. Cote 2003.

26. Gemeente Batavia 1937, 70.

27. Colonial scientists published the results of laboratory tests on various sources of groundwater and surface water, giving readers detailed biochemical content of each source and including an illustrated appendix with diagrams of the microscopic organisms (Maronier 1929, 292, 230).

28. Maronier 1929, 225.

29. See Milone 1967; Van der Kroef 1954; Van Doorne 1983.

30. Van der Kop 1926, 149 (emphasis added).

31. Van Leeuwen 1920, 198; Vervoort 1926, 266.

32. Van Breen 1919, 138.

33. Vervoort 1926, 266.

34. Abeyasekere 1989.

35. Van Doorne 1983.

36. Mrazek 2002.

37. Argo 1999; Maronier 1929.

38. Van Raay 1915.

39. Karsten 1958, 42.

40. A spring-water supply system delivering pressurized, potable water into the city was completed in 1922 (Smitt 1922).

41. Van Leeuwen 1917.

42. Kusno 1997.

43. Hanna 1959; Kusno 1997.

44. DKI 1972, 85.

45. In particular, two major water-treatment plants were completed in the 1950s and 1960s, adding three thousand liters to Jakarta's water production capacity.

46. Production capacity systematically exceeded distribution capacity in the decades following independence (Martijn 2005).

47. Kusno 2000.

48. Kusno 1997; Leclerc 1993.

49. Abeyasekere 1985.

50. Chifos and Hendropranoto 2000; Firman and Dharmapatni 1994; Ford 1993; Silas 1992; World Bank 1999, 2004.

51. Fischer 1959; Kusno 2000.

52. In 1948, prior to official independence, Jakarta's population was 823,000; by 1952 the population was estimated at 1.8 million, and by 1965 it was 3.8 million (Cowherd 2002). For figures on connection rates over time, see Fischer 1959;

Hanna 1959; PAM Jaya 1992a; Azdan 2001; Cessti 1993; FORKAMI 2006; JICA 1997; Porter 1996; Sabarini 2007.

53. The remainder was taken up by industrial and government consumption (Azdan 2001; JICA 1997).

54. KIP 1976.

55. Taylor 1983.

56. The World Bank–financed PAM Jaya Supply Improvement Project, 1990–1997. Although this project resulted in over two hundred thousand new connections, the expansion of production and distribution capacity did not catch up to increasing (potential) demand due to population growth (Cessti 1993; JICA 1997; PAM Jaya 1992a; PAM Jaya 1992b; Porter 1996; World Bank 1998).

57. Official development plans for Jakarta encouraged an east-west pattern of urban development, attempting to avoid expansion into irrigated agricultural areas north and south of the city (JICA 1997).

58. Woodcock 2006.

59. Argo 1999, 71.

60. Crane 1994; Pandjaitan 2004.

61. For water supply utilities (PDAMS) across Indonesia, the standard level of this dividend is 55 percent of net profits, which is paid to the treasury of the local municipality (World Bank 1997b). However, this is mandatory only when 75 percent coverage of the population is reached (for PDAMs in urban areas); moreover, in practice, all or part of this dividend is frequently returned to the utility. (See also PAM Jaya 1992b; Taylor 1983). The method of "user fees" to finance network extension meant that house connections were unaffordable for the majority of the population; the cost of a household connection (not to mention "additional fees," meter rental, deposit, and actual monthly tariffs) in 1975 was Rp 100,000 (US$200), whereas the average income in Jakarta at that time was only Rp 15,000/month (approximately US$36), with the 80 percent of the city's residents that lived in kampungs earning much less than that amount (DKI Jakarta 1976).

62. Banded tariff structures with a rising block tariff, beginning with rates below production cost, create a disincentive for providing direct network connections to poor customers, who pay lower amounts per unit volume; large numbers of poor customers thus threaten to decrease water company revenues and could theoretically result in revenue per unit volume falling below marginal cost (Whittington 1992).

63. JICA 1997.

64. The treatment of water utilities as "cash cows" (notably through the payment of annual dividends to the municipality) has in some cases distorted long-term planning processes, reducing investment in infrastructure maintenance and renewal and reducing funds available for network expansion (World Bank 1997b).

65. Martijn 2005.

66. Ibid.

67. Porter 1996; JICA 1997.

68. DKI Jakarta (Daerah Khusus Ibukota [Special Capital District of Jakarta]) alone grew from an estimated 1.8 million people in 1950 to an estimated 6.5 million in 1980, with equally rapid population growth in the surrounding metropolitan areas, particularly at the expanding rural-urban fringe beyond the boundaries of DKI Jakarta (Chifos 2000; Firman 1998, 1999, 2004; Lo and Yeung 1996). The most notable attempt to extend water supply access to the poor was via the Kampung Improvement Program (KIP) (1966–77), but this produced only marginal improvements in the access of lower-income residents to piped water (Abeyasekere 1985; KIP 1976; Taylor 1983; World Bank 1974). On increased industrial supply, see PAM Jaya 1992c.

69. Ohlsson 2000.

70. Darrow 2005, 475.

71. See UNDP 2005; World Bank 2006b.

72. See Bakker et al. 2008. For a history of the preferences of poor households for alternative water supplies, see Kreimer et al. 1995; Surjadi et al. 1994; Yayasan Dian Desa 1990; Chifos 2000; Chifos and Hendropranoto 2000.

73. Our evidence is drawn from primary survey data based on a survey of 110 households in six Jakarta neighborhoods in 2005, and is supported by references to other published survey data. The survey controlled for gender bias by selecting only female respondents (on the basis of the significantly higher responsibility of women for all water-related activities within the home). The average household monthly income of Rp 1.4 million (with an average of 5.7 household residents) translates into an average individual income of Rp 245,614, slightly above the 2005 official poverty line but well below the level of income earned by minimum-wage earners. This method ensured a uniform, comprehensive, and randomized approach to sample selection. But it should be noted that the deliberate choice of communities with self-identified problems of water access was intended to document the governance failures affecting poor households in areas where some network supply (and hence at least the theoretical possibility of connecting to the network) did exist, rather than being representative of all households across Jakarta, many of which (both poor and wealthy) are in areas completely without water network access. For more details on the survey, see Bakker et al. 2008.

74. Jellinek 1991; Jellinek 1997.

75. Kjellén 2006.

76. Aman Tirta and USAID 2006.

77. Surjadi et al. 1994.

78. ESP 2006; Mukherjee 2006.

79. See Mujianto 2004; Mukherjee 2006.

80. For a poor household where residents consume fifty liters/person/day (the World Health Organization recommended minimum), fixed charges will be anywhere from five to ten times as high as the volumetric consumption charge.

81. See Cowherd 2002 for a discussion of the culture of "informal" profits from public services; also Shofiani 2003; Susantono 2001; Yayasan Dian Desa 1990.

82. Jellinek 1991.

83. Harsanto 2005.

84. FORKAMI 2007; Harsanto and Wahyudi 2002.

85. FORKAMI 2007; Nurbianto 2007; Sabarini 2007.

86. From approximately Rp 2,300/US$1 in 1997 to Rp 10,000/US$1 in 1998.

87. The private companies were associated with Bob Hassan, a close friend of the then-president Suharto, and with Sigit Harjojudanto, the president's eldest son. Note that corruption in Indonesia under Suharto was intertwined with political governance, and hence popularly labeled *"korupsi, kolusi, dan nepotisme"* (KKN, or "corruption, collusion, nepotism"), which came to symbolize the Suharto regime (Robertson-Snape 1999; Scott 1969). That the contracts were awarded despite national laws prohibiting foreign investment in drinking water delivery (Law no. 1/1967; Ministry of Home Affairs Decision no. 3/1990) and local regulations (no. 11/1992 and no. 11/1993) precluding private sector involvement in community drinking-water supply was to be a source of conflict in the early years of the contract (Argo and Firman 2002).

88. Harsoso 2005.

89. The cumulative deficit by the end of 2001 was Rp 469 billion (approximately US$46 million) and had reached Rp 990 billion (approximately US$97 million) by September 2003—excluding late payment interest and retroactive tariff increases (Jakarta Water Supply Regulatory Body 2005a, 2005b).

90. By 2002, the service coverage level for the two concession areas remained well below the 70 percent target specified for 2002 in the initial contract (Platts 2002).

91. The latest figures available from Palyja (2009 data), the private company covering the western half of Jakarta, indicate a service coverage of 63.93 percent, with a total of 412,456 connections (PALYJA 2009). See also JWSRB 2008.

92. The lowest tariff in 2003 was Rp 550/cubic meter, well below the production cost. A secondary disincentive is the higher average cost per connection in poor neighborhoods, which raises installation costs. Given the lack of land-use planning in informal settlements, the highly dense and disordered distribution of homes means that installing connections may be more time consuming (if conventional below-ground infrastructure is used). For a public operator, these disincentives were direct. For private operators, they were indirect, but just as important, because the revenue received by the municipality is the source of funds from which the private operators are paid, and decreasing revenues implies a greater chance of debt, longer repayment period, and increased possibility of government default on debts now held by the private companies—in addition to the multilateral lending organizations (most importantly the World Bank) which had funded water supply loans prior to the 1990s.

93. Bakker 2007b; BPD 2001.

94. Forkami 2006.

95. Here, I am referring to an "output-based aid" project, initiated in Jakarta through the World Bank's Global Partnership on Output-Based Aid. Intended to "accelerate" the connection of poor households (given that the two private firms had failed to meet the original contract targets for new connections), the project had resulted in late 2008 in only four thousand of an original target of forty thousand new connections.

96. The precise figures are from 108,159 to 159,493 "upper-class households and small business," and from 64,829 to 116,649 "middle-class" households (PALYJA 2008).

97. In 2006, Thames Water sold its stake in its Jakarta water services subsidiary to an Indonesian-led consortium (AERTRA), which now manages the eastern half of the city. Palyja succeeded in salvaging a proportion of its investments through floating a domestic bond issue for water investment; it continues to operate the management contract for the western half of the city.

98. Chatterjee 2004, 50.

99. Bond 2003.

5. Protesting Privatization

1. Winpenny 2003.

2. Only two of the United Nations conventions on human rights (article 24 of the Convention on the Rights of the Child, and article 14 of the Convention on the Elimination of All Forms of Discrimination against Women) explicitly recognize the right to water. No explicit right to water is expressed in the most relevant international treaty, the International Covenant on Economic, Social and Cultural Rights, one of the keystones of international human rights law (Morgan 2004a).

3. In Uruguay and South Africa, the right to drinking water is enshrined in the constitution of each country. In France, the right to water is established in national legislation. For a global survey, see Smets 2006b.

4. For a survey of the human right to water in national legislation, see Smets 2006b. For reviews of the human right to water, focusing on questions of implementation, see COHRE 2008 and WHO 2003b.

5. Reasons suggested for the emergence of a rights-based approach include the impact of structural adjustment programs, the failure of "basic needs" approaches to development, the end of the cold war, and the demise of authoritarian regimes in some countries. For a discussion of rights-based approaches to development, see Gready and Ensor (2005), and Darrow (2005).

6. Alter-globalization proponents oppose what they term "corporate globalization" and are broadly supportive of alternative models of globalization: democratic, transparent, locally accountable, equitable, and (in some variants) anticapitalist. See, for example, Brecher, Costello, and Smith (2000).

7. Glassman 2006; Harvey 2003a.

8. Assies 2003; Barlow and Clarke 2003; Hukka and Katko 2003; McDonald and Ruiters 2005; Petrella 2001; Shiva 2002.

9. See, for example, Bakker 2004; Allouche and Finger 2001; Johnstone and Wood 2001; Kloezen 1998; Laurie and Marvin 1999; Swyngedouw 2005.

10. See, for example, Bailey 2005; Barrett and Jaichand 2007; Bond and Dugard 2008; Johnston, Gismondi, and Goodman 2006; Laxer and Soron, 2006; McDonald and Ruiters 2005.

11. Johnstone and Wood 2001; Kumar and Singh 2001; Rogers and Hall 2003; Shirley 2002; Winpenny 1994.

12. Similar arguments have been made with respect to energy use and "fuel poverty." See Boardman 1991, 1999; DTI 2001.

13. European Parliament 2000.

14. UN ECOSOC 2003.

15. Bauer 1997.

16. Loftus and McDonald 2001; Perreault 2006.

17. Hall, Lobina, and de la Motte 2005.

18. Giarracca and del Pozo 2005; Holland 2005; Morgan 2004b.

19. See also Bakker 2004.

20. Hall, Lobina, and de la Motte 2005.

21. Khagram 2004.

22. Conca 2006.

23. Khagram 2004.

24. Gandy 2002; Haddad 2000.

25. Bauer 1998; Haddad 2000.

26. Bakker 2005.

27. Gleick 1998; Hukka and Katko 2003; Morgan 2004b, 2005; Trawick 2003.

28. These declarations include the Cochabamba Declaration, the Group of Lisbon's Water Manifesto (Petrella 2001), and the Declaration of the P8 (the world's poorest eight countries, organized as a counterpart to the G8) at their fourth summit in 2000. Campaigns include the UK-based Right to Water (http://www.righttowater.org.uk), the Octubre Azul/Blue October, the Canada-based Friends of the Right to Water, the U.S.-based Water for All, and the Green Cross campaign for an international convention on the right to water (http://www.watertreaty.org).

29. The Water Supply and Sanitation Collaborative Council, located in Geneva, is a nonprofit organization that acts as an "international policy think tank" on water management.

30. The Cochabamba Declaration followed a meeting of several hundred people in this Bolivian city concerned about the involvement of private sector corporations in water supply management. See http://www.canadians.org/blue planet/cochabamba-e.html.

31. The Group of Lisbon is a group of distinguished scholars from around the world who analyze globalization and call for new types of economic governance. See Petrella 2001.

32. Chairperson of the Council of Canadians (one of the country's largest civil-society organizations) and joint winner of the 2005 Right Livelihood Award for her work on water, Ms. Barlow had been campaigning against water privatization for over a decade at the time of her appointment.

33. ECOSOC 2002; WHO 2003b; UNDP 2006.

34. The Committee on Economic, Social and Cultural Rights (CESCR) is the body of independent experts that monitors implementation of the International Covenant on Economic, Social and Cultural Rights by its "states parties." See ECOSOC 2002; Hammer 2004.

35. Specifically articles 11 and 12. The covenant has been signed and ratified by 145 states, and signed but not ratified by seven states, including the United States (UN-OHCHR 2002).

36. CESCR 2002. See also related reports of the Office of the United Nations High Commissioner for Human Rights on the human right to water. http://www2.ohchr.org/english/issues/water/.

37. The Human Rights Council is an intergovernmental body within the UN system made up of forty-seven states responsible for strengthening the promotion and protection of human rights. The council was created by the UN General Assembly in 2006 with a mandate of addressing situations of human-rights violations and making recommendations on them.

38. UNWWAP 2006.

39. UNDP 2006.

40. Vandenhole and Wielders 2008.

41. Alston 2005.

42. Brooks 2008.

43. WHO 2003a. Gleick (1996) recommends 50 liters per person per day.

44. Dennis and Stewart 2004.

45. Van Harten 2005.

46. Schreiber 2008.

47. Bond 2002; McDonald 2002a; McDonald and Ruiters 2005.

48. UNWWAP 2006.

49. For example, under Islam, water is defined as collective property (*waqf*), with water available free in specific cases (Faruqui, Biswas, and Bino 2001).

50. Bluemel 2004.

51. See, for example, Boelens 2008, 2009.

52. Boelens and Hoogendam 2002; Derman and Hellum 2007; Zwarteveen 1997.

53. Boelens 2009.

54. Langford 2005.

55. Barrett and Jaichand 2007; Langford 2006.

56. For example, Veolia, one of the largest private water companies in the world, states on its French-language website: "L'eau est considérée à la fois comme un bien économique, social, écologique et comme un droit humain." (Water is considered an economic, social, and ecological good as well as a human right.) See http://www.veoliaeau.com/gestion-durable/gestion-durable/eau-pour-tous/bien-commun. See also the comments of Antoine Frérot (at the time, director general of Veolia) during the Open Forum on "Water: Property or Human Right?" at the 2004 Davos Forum, at http://gaia.unit.net/wef/worldeconomicforum_annualmeeting2006/default.aspx?sn=15810.

57. Bailey 2005; Segerfeldt 2005.

58. Dubreuil 2006a, 2006b.

59. Salman and McInerney-Lankford 2004.

60. Payen 2006.

61. Smets 2006a.

62. Robbins 2003.

63. UNDP 2003; ADB 2003a; IMF 2004.

64. As recognized by the UN committee in its general comment on the human right to water, which observes that, in permitting third parties (such as the private sector) in addition to state actors to supply water, an additional burden is placed on regulatory frameworks, including "independent monitoring, genuine public participation, and imposition of penalties for non-compliance" (ECOSOC 2002, article 24).

65. Kneen 2009.

66. The Constitution of the Republic of South Africa guarantees the right of citizens to access "sufficient water" and places an onus on the state to "take reasonable legislative and other measures, within its available resources, to achieve the progressive realization" of this right (Act 108 of 1996, section 27[2]).

67. DWAF 1995.

68. Hemson 2004.

69. Galvin and Habib 2003.

70. Under the terms of the Water Services Act (1997) and National Water Bill (1998). National government retains control of water resources management.

71. Van der Merwe 1995.

72. Prepayment meters were banned by Britain's High Court in 1997, following legal challenges by municipalities. See Bakker 2004; Bond and McInnes 2007.

73. DWAF 1998.

74. Bond 2007.

75. Langa 1997.

76. Government of South Africa, *Mid-Year Population Estimates* (Pretoria: Statistics South Africa, 2007).

77. Bond 2003.

78. Hemson and Dube 2004.

79. Mugero and Hoque 2001.

80. Bond 2002; McDonald 2002b.

81. Tissington et al. 2008.

82. Johannesburg Water is a wholly owned corporatized water utility, which in 2001 entered into a five-year management contract with an international consortium headed by French water multinational Suez. At the end of the contract, the city's water was returned to municipal management.

83. For copies of all relevant court documents, see http://www.law.wits.ac.za/cals/phiri/index.htm.

84. City of Johannesburg 2008.

85. South Africa's National Water Act specifies the "right of all citizens to have access to basic water services (the provision of potable water supply and the removal and disposal of human excreta and waste water) necessary to afford them a healthy environment on an equitable and economically and environmentally sustainable basis" (National Water Act, Act 36, 1998, Parliament of South Africa).

86. City of Johannesburg Answering Affidavit Part 1, case no. 06/16835, High Court of South Africa (Witwatersrand Local Division).

87. The ruling of fifty liters per person per day was an increase from the current allocation whereby each household (on average containing sixteen persons) was only provided with two hundred liters per day.

88. See, for example, Olivera and Lewis 2004; Shiva 2002.

89. Langford 2005.

90. See Jaglin 2002.

91. Darrow 2005.

92. Morgan 2008.

93. See Morgan 2008; UNDP 2000.

94. Mitchell 2003. See also Harvey 2003b.

95. See Alvarez, Dagnino, and Escobar 1998.

96. See Taylor (2004) for a discussion of the origins of this dichotomy (in European societies), expressed in the seventeenth and eighteenth centuries as a tension between new economic actors and the warrior ethos of the nobility.

6. Commons versus Commodities

1. On the water commons, see Barlow 2007. On water democracy, see Shiva 2002. On remunicipalization, see Balanyá et al. 2005. See also Petrella 2001 and the "remunicipalisation tracker" cosponsored by the Transnational Institute and Corporate Europe Observatory at http://www.remunicipalisation.org.

2. See Weitz and Franceys 2002; Narain 2006; UNDP 2006; UNWWAP 2006.

3. See, for example, Barnes 2006, and (for a critical reading) McCarthy 2005.

4. Rowe 2008, 150.

5. Wade 1988.

6. See, for example, Barlow 2007 and Olivera n.d.

7. The following account is based on first-hand interviews conducted in Cochabamba in 2001. See also the following sources: Assies 2003; Bustamante, Peredo, and Udaeta 2005; Olivera and Lewis 2004.

8. Only 55 percent of urban and 46 percent of rural residents have access to water-supply and sanitation networks in Cochabamba (Bustamante, Peredo, and Udaeta 2005).

9. Laurie and Marvin 1999.

10. García, García, and Quiton 2003.

11. Bustamante, Peredo, and Udaeta 2005; Olivera and Lewis 2004; Perreault 2006; Schultz 2008; Spronk and Flores 2008.

12. Boelens and de Vos 2006.

13. Schultz 2008.

14. Calle and Lohman 2006.

15. Chatterjee 2004.

16. These members might be defined by geographical location (as with the British village "commoners"), by purchase of shares (as with cooperative banks), or by affiliation (as with contemporary proposals to create national public "trusts" of which all citizens are members).

17. See Ostrom 1990; Ostrom and Keohane 1995; Ostrom, Schroeder, and Wynne 1993; Roberts and Emel 1992.

18. Rivera 1998.

19. Ostrom and others make a distinction between common-pool resources (rivalrous, nonexcludable, subtractible) and public goods (nonrivalrous, nonexcludable, nonsubtractible). In both cases, it is difficult to exclude people from access. But with common-pool resources, such as a fishery, use by one person can reduce the overall resource. Public goods (such as national defense or street lighting) are not "diminishable" in the same manner (Ostrom 1990).

20. Hardin 1968.

21. Berkes 1989; Feeny, Hanna, and McEvoy 1996.

22. Caffentzis 2004.

23. Ostrom 1990.

24. Bacdayan 1974; Mosse 2003; Trawick 2001b; Wade 1988.

25. Heikkila 2004; Lopez-Gunn 2003; Marshall 2004; Wade 1988.

26. Cleaver 2000; Mehta 2001; Mosse 1997; Ostrom 1990; Ostrom and Keohane 1995; Potkanski and Adams 1998; St. Martin 2005.

27. Ostrom 1990.

28. I am grateful to Rutgerd Boelens for this insight.

29. Ashenafi and Leader-Williams 2005; Boelens and Doornbos 2001; Cremers, Ooijevaar, and Boelens 2005; Potkanski and Adams 1998; Sokile and van Koppen 2004.

30. Strang 2004.

31. Mosse 2003.

32. Ley número 2878, de promoción y fomento al sector riego.

33. García, García, and Quiton 2003; Olivera and Lewis 2004; Perreault 2006.

34. Boelens and de Vos 2006; Boelens and Zwarteveen 2005; de Vos, Boelens, and Bustamante 2006; Perreault 2005.

35. Finnegan 2002; Nickson and Vargas 2002; Perreault 2005; Perreault 2006.

36. Aldurralde et al. 2005.

37. I am grateful to Rocío Bustamante for these insights.

38. Chile encountered some of these same issues after it created a system of water rights. See Bauer 1997, 1998; Budds 2004.

39. See, for example, Sabatier et al. 2005.

40. Agnew 1999; Pierre and Peters 2000; Rhodes 1996; Swyngedouw 1997, 2004.

41. Kemper, Dinar, and Blomquist 2005.

42. Akbar et al. 2007; Gaye and Diallo 1997; Kyessi 2005; Sabatier et al. 2005.

43. A cooperative may be simply defined as an enterprise owned and democratically controlled by the users of the goods and services provided; users can be consumers, employees, or producers of products and services.

44. Juuti, Katko, and Rajala 2005; Katko 1992.

45. Birchall 2001; Birchall and Simmons 2004; Birchall 2002; Kay 1996.

46. Bakker 2003b; Co-operatives Secretariat 2001; Curry and McGuire 2002; Emel and Roberts 1995.

47. Hall, Lobina, and de la Motte 2005; Ruiz-Meir and van Ginneken 2006; Yavari 2005.

48. The Welsh case is of particular interest as it highlights the potential advantages of cooperatives, mutuals, and other similar not-for-profit, customer-directed corporate governance structures (Bakker 2003b). The joint USAID-JBIC "Clean Water for People Initiative" has built on this model.

49. For a discussion pertaining to water, see Norman and Bakker 2009. For a more general discussion, see Cooke and Kothari 2001.

50. Jaglin 2002.

51. Boelens 2008.

52. Norman and Bakker 2009.

53. Balanyá et al. 2005.

54. Kay 1996.

55. See Blokland, Braadbaart, and Schwartz 1999.

56. Lemos and De Oliveira 2005; Muzzini 2005; Ugaz 2002.

57. See Scottish Parliament 2001. The failure of Bolivia's Santa Cruz water-supply cooperative to expand beyond the "fifth ring" of the city is a good example, as is the cooperative's coverage rates for sanitation sewerage (approximately 50 percent), causing considerable groundwater contamination (Spronk 2007).

58. Cleaver and Toner 2006.

59. Indeed, fears about capture of boards by "sectional" interests were a key factor in the decision by the British water industry's economic regulator against tentative proposals by British water companies to "mutualize" their operations following the successful mutualization of Welsh Water. See Bakker 2003b. See also Platteau and Gaspart 2003.

60. Cooke and Kothari 2001.

61. Weitz and Franceys 2002; Mugabi, Kayaga, and Smout 2007; Stewart and Gray 2006.

62. See, for example, the World Bank–sponsored Business Partners for Development; the UK government–sponsored Partners for Water and Sanitation network (http://www.partnersforwater.org); and the EU-sponsored Water Initiative (http://www.euwi.net).

63. Hall and Lobina 2007; Jaglin 2002.

64. Collignon and Vezina 2000; World Bank 2003a.

65. Activists have energetically promoted these public-public-partnerships, and the UN Secretary-General's Advisory Board on Water and Sanitation (UNSGAB) has sponsored a formal initiative. Encouraged by the UN Commission on Sustainable Development's official acknowledgment of the importance of promoting public-public partnerships, and by specific campaigns by public water-supply utilities—notably in Porto Alegre—governments in Argentina, Bolivia, Brazil, Indonesia, Holland, Honduras, France, South Africa, and Sweden have initiated public-public partnerships (Hall and Lobina 2006b, 2007; PSIRU 2006; Public Citizen 2002; Balanyá et al. 2005; UNCSD 2005). There are other examples, such as the international extension by the U.S. National Rural Water Association of its "circuit rider" program for training small-system operators (http://www.nrwa.org/internationalruralwater/index.html).

66. Bennholdt-Thomsen and Mies 2000; Meillassoux 1981; Perelman 2000.

67. Strang 2004.

68. Cantin-Cumyn 2006–2007.

69. Corbridge et al. 2005

70. McCarthy 2005; Mehta 2001; Mehta, Leach, and Scoones 2001.

71. See Baland and Platteau 1997, 1998.

72. I am indebted to Isha Ray for this insight.

73. McCarthy 2005.

74. For example, in order to be financially sustainable, community-run systems often focus on (relatively) wealthier customers or neighborhoods, and/or exclude households that pose a risk of nonpayment. Other risk factors (such as high mobility) also reduce the chances of the most disadvantaged households gaining access to water. This illustrates the limited degree to which the urban poor are able to address issues of distributive justice; on the contrary, precisely because of their financial precariousness, they are likely to be driven to exclude other, more disadvantaged users.

75. Birchall, for example, cautions that there are many other variables at work, citing cases where a transfer of ownership of assets to a nonprofit corporation or cooperative has not had a noticeable effect on governance (Birchall 2001).

76. For instance, consider the degree to which consumer representatives can adequately and accurately reflect the interests of different constituencies. Other frequently voiced concerns include the likelihood of regulatory capture and information asymmetries between consumers, the regulator, and the water supplier, not to mention the high costs of ensuring transparency and facilitating broad-based participation. Garande and Dagg 2005; Wolff and Hallstein 2005.

77. This is another way of formulating the argument (discussed in earlier chapters) that ownership (i.e., public versus private) is less important than institutions (rules, norms, and laws) and governance (decision-making processes). There are many potential reasons for this. Formal governance mechanisms may be less important than the way in which governance mechanisms work in practice; democratic decision-making processes may not necessarily imply improved efficiency, accountability, or managerial outcomes. This may be the case because the degree of power that consumers can effectively assert does not flow automatically from consumer participation in utility regulation. Other important barriers are the scope of representation (and the degree to which consumer representatives can adequately and accurately reflect the interests of different constituencies); the need for independent consumer representatives (despite the threat of regulatory capture or capture by sectional interests); the likelihood of information asymmetries between consumers, the regulator, and the water supplier; and the high costs of ensuring transparency and facilitating broad-based participation. See Garande and Dagg 2005; Wolff and Hallstein 2005.

78. Balanyá et al. 2005.

79. Kay 1996.

80. See Blokland, Braadbaart, and Schwartz 2001.

81. Bakker 2004.

82. Castro 2007.

83. Chatterjee 2004.

84. Corbridge et al. 2005.

85. For example, Appadurai 2002.

86. This, of course, is consistent with the flourishing of the "new localism," which views local involvement as necessary and positive, both as a means of supplanting higher-order levels of government and as a mechanism to foster social capital. Brown and Purcell 2005, 607; Corry et al. 2004; Raco 2000; Raco and Flint 2001.

87. Corry et al. 2004; De Loe, Di Giantomasso, and Kreutzwiser 2002; Gibbins 2001; O'Riordan 2004.

88. Cochrane 1986; Evans 2004.

89. The local trap is an analytical counterpart to Agnew's "territorial trap," which was originally applied to debates within international relations and political geography but is also debated within the environmental governance literature. See Norman and Bakker 2009.

90. Gibson-Graham 2006.

91. Fischhendler and Feitelson 2005; Gleick 1993; Kliot, Shmueli, and Shamir 2001.

92. De Loe, DiGiantomasso, and Kreutzwiser 2002.

93. Cochrane 1986; Taylor 2004; Van Rooy 1997, 2004.

7. POLITICS AND BIOPOLITICS

1. Kooy and Bakker 2008a, 2008b.

2. Bashford 2004.

3. Goubert 1989.

4. Gandy 2004, 2006; Kaika 2005.

5. See, for example, Gandy 1999, 2002; Kaika 2005.

6. See Guha (2000) for a summary of debates over the "ecology of the poor."

7. See Bakker 2003c; Castro, Kaika, and Swyngedouw 2002; Swyngedouw 1997.

8. The "new culture of water" paradigm, which originated in Spain in the mid-1990s in response to a planned wave of megadams and hydraulic projects, is one attempt. The main force behind the popularization of the paradigm is Pedro Arrojo, a Spanish academic and winner of the Goldman Environmental Prize. "Integrated Water Resources Management" is another, although slightly narrower, formulation of similar ideas. See Arrojo 2006; Biswas 2008; Mitchell 2005.

9. There is a range of works within the field of political ecology that develop similar perspectives. For a discussion of "ecological governance," see Gale and M'Gonigle 2000.

10. Harvey 2005.

11. Bernstein 2001.

12. O'Connor 1996.

13. Katz 1998.

14. Perelman 2000.

15. Jasanoff 2004; Latour 1993; McKibben 1989; Swyngedouw 1999.

16. Beck 1992.

17. Polanyi 1944.

18. Bakker 2005.

19. Ibid.

20. Harvey 2003a.

21. See Shiva 2002.

22. Arrojo 2006. For the text of the European Declaration for a New Culture of Water, see http://www.unizar.es/fnca/euwater/index2.php?idioma=en.

23. The public-trust doctrine has received its fullest contemporary judicial expression in the United States but forms part of the legal traditions of many countries.

24. Kaika 2005.

25. Distributive justice is centrally concerned with the way in which "property is distributed throughout society so that everyone is supplied with a certain level of material means" (Fleishacker 2004, 4).

26. E.g., Boardman 1991, 1999.

27. Gandy 1999, 2004; Kaika 2005.

28. Solley, Pierce, and Perlman 1998.

29. WHO/UNICEF 2006.

30. See UNDP 2006.

31. See Shiva 2002.

32. R. Bustamante, "Debemos comenzar a cuestionar la idea del derecho al agua" (seminar presentation, Modelos de gestión de agua en ciudades y comunidades de los Andes, La Paz, 2005), http://www.pieb.com.bo/noticia.php?idn=2055.

33. Boelens 2008.

34. There have been similar arguments in the past: Wittfogel's much-criticized characterization of "hydraulic despotism" is a classic example. See Worster 1985.

CONCLUSION

1. Some of this information is drawn from interviews conducted with staff of the Porto Alegre water department (DMAE) in 2003. See also Hall et al. 2002; Heller 2001; Menegat 2002; Balanyá et al. 2005.

2. These services have included activities considered "core" to the business, including water and sewage connection services, transportation services, machinery and equipment rental, vehicle rental, and data processing.

3. Guha 2000.

4. Barlow 2007.

5. Caffentzis 2004.

6. McCarthy 2005.

7. Ostrom 1990.

8. See Heller 2001.

9. I am indebted to Patrick Bond, Laila Smith, and COHRE for discussions on the South African case, as well as to the staff of South African NGO Mvula Trust, with whom I spent several months in 1998. This example was inspired by the discussion in Bond et al. (2008) of the "neoliberal loo." See also Bond 2002; Muller 2007, 2008; and, for a rebuttal, Bond and Dugard 2008.

10. Bond and Dugard 2008.

11. D. Hall, *Economic Crisis and Public Services Note no. 1*. Greenwich: Public Services International Research Unit, University of Greenwich, 2008. The Stiglitz report on the costs of the Iraq War is perhaps the most reliable. See http://www.nytimes.com/2009/03/01/weekinreview/01glanz.html.

12. Altenburg and von Drachenfels 2006.

REFERENCES

Abeyasekere, S. "Colonial Urban Politics: the Municipal Council of Batavia." *Kabar Seberang* 13, 14 (1984): 17–24.

——, ed. *From Batavia to Jakarta: Indonesia's Capital 1930s to 1980s*. Melbourne: Monash University, 1985, 1–23.

——. "Death and Disease in 19th Century Batavia." In *Death and Disease in Southeast Asia: Explorations in Social, Medical, and Demographic History*, edited by N. Owen. Oxford: Oxford University Press, 1987.

——. *Jakarta: A History*. Singapore: Oxford University Press, 1989.

ActionAid. *Turning Off the Taps: Donor Conditionality and Water Privatisation in Dares Salaam, Tanzania*. Report. London: ActionAid International, 2004.

ADB, Asian Development Bank. "Beyond Boundaries: Extending Services to the Urban Poor." Manila: Asian Development Bank, 2003a.

——. "Water in Asian Cities." Manila: Asian Development Bank, 2003b.

Agarwal, A. "Common Resources and Institutional Sustainability." In *The Drama of the Commons*, edited by O. Elinor, T. Dietz, N. Dolsak, P. Stern, S. Stonich, and E. Weber, 41–86. Washington, DC: National Academies Press, 2002.

Agnew, J. "Regions on the Mind Does Not Equal Regions of the Mind." *Progress in Human Geography* 23, no. 1 (1999): 91–96.

Agtini, M., R. Soeharno, M. Lesmana, N. Punjabi, C. Simanjuntak, F. Wangsasaputra, D. Nurdin, et al. "The Burden of Diarrhoea, Shigellosis, and Cholera in North Jakarta, Indonesia: Findings from 24 Months Surveillance." *Infectious Disease* 1, no. 5 (2005): 89–100.

Ahmed, S. "Water Resources and the Issue of Social Justice: The Capability Approach." PhD diss., Northern Arizona University, 2008.

Akbar, H., J. Minnery, B. van Horen, and P. Smith. "Community Water Supply for the Urban Poor in Developing Countries: The Case of Dhaka, Bangladesh." *Habitat International* 31, no. 1 (2007): 24–35.

Alberta Environment. *Water for Life: Alberta's Strategy for Sustainability*. Edmonton: Alberta Environment, 2003.

Aldurralde, J., J. Molina, E. Vilarroel, and R. Orrelana. *Informe final: Proyecto "Regulación de derechos de agua en Bolivia."* Final Report 101423. Ottawa: International Development Research Centre, 2005. http://idl-bnc.idrc.ca/dspace/handle/123456789/29576.

Allouche, J., and M. Finger. *Water Privatisation: Trans-national Corporations and the Re-Regulation of the Water Industry.* London: Taylor and Francis, 2001.

Alsayyad, N., and A. Roy. "Medieval Modernity: On Citizenship and Urbanism in a Global Era." *Space and Polity* 10, no. 1 (2006): 1–20.

Alston, P. "Ships Passing in the Night: The Current State of the Human Rights and Development Debate Seen through the Lens of the Millenium Development Goals." *Human Rights Quarterly* 27, no. 3 (2005): 755–829.

Altenburg, T., and C. von Drachenfels. "The 'New Minimalist Approach' to Private-Sector Development: A Critical Assessment." *Development Policy Review* 24, no. 4 (2006): 387–411.

Alvarez, S., E. Dagnino, and A. Escobar. *Cultures of Politics, Politics of Cultures: Re-visioning Latin American Social Movements.* Boulder, CO: Westview Press, 1998.

Aman Tirta, and USAID. *Air Rahmat: Air minum sehat untuk Indonesia.* Jakarta, 2006.

Amin, A., and S. Graham. "The Ordinary City." *Transactions of the Institute of British Geographers* 22, no. 4 (1997): 411–429.

Anderson, T., and D. Leal. *Free Market Environmentalism.* New York: Palgrave Macmillan, 2001.

Annez, P. "Urban Infrastructure Finance from Private Operators: What Have We Learned from Recent Experience?" World Bank Policy Research Working Paper 4045. Washington, DC: World Bank, 2006.

Appadurai, A. "Deep Democracy: Urban Governmentality and the Horizon of Politics." *Public Culture* 14, no. 1 (2002): 21–47.

Arendt, H. *The Human Condition.* Chicago: University of Chicago Press, 1958.

Argo, T. "Thirsty Downstream: The Provision of Clean Water in Jakarta, Indonesia." PhD diss., University of British Columbia, 1999.

Argo, T., and T. Firman. "To Privatize or Not to Privatize? Reform of Urban Water Supply Services in Jabotabek, Indonesia." *Built Environment* 27, no. 2 (2002): 146–155.

Arrojo, P. *El reto ético de la nueva cultura del agua: Funciones, valores y derechos en juego.* Madrid: Paidós Iberica, 2006.

Ashenafi, Z.T., and N. Leader-Williams. "Indigenous Common Property Resource Management in the Central Highlands of Ethiopia." *Human Ecology* 33 (2005): 539–563.

Assies, W. "David versus Goliath in Cochabamba: Water Rights, Neoliberalism, and the Revival of Social Protest in Bolivia." *Latin American Perspectives* 30, no. 3 (2003): 14–36.

Azdan, M. "Water Policy Reform in Jakarta, Indonesia: A CGE Analysis." PhD diss., Ohio State University, 2001.

Bacdayan, A. "Securing Water for Drying Rice Terraces: Irrigation, Community Organization, and Expanding Social Relationships in a Western Bontoo Group, Philippines." *Ethnology* 13, no. 3 (1974): 247–260.

Baietti, A., W. Kingdom, and M. van Ginneken. "Characteristics of Well-Performing Public Water Utilities." *Water Supply & Sanitation Working Notes,* no. 9. Washington, DC: World Bank, 2006.

Bailey, R. "Water Is a Human Right: How Privatization Gets Water to the Poor." *Reason Online,* August 17, 2005.

Bakker, K. "Archipelagos and Networks: Urbanization and Water Privatization in the South." *Geographical Journal* 169, no. 4 (2003a): 328–341.

——. "From Public to Private to...Mutual? Restructuring Water Supply Governance in England and Wales." *Geoforum* 34, no. 3 (2003b): 359–74.

——. "A Political Ecology of Water Privatization." *Studies in Political Economy* 70 (2003c): 35–58.

——. *An Uncooperative Commodity: Privatizing Water in England and Wales*. Oxford: Oxford University Press, 2004.

——. "Neoliberalizing Nature? Market Environmentalism in Water Supply in England and Wales." *Annals of the Association of American Geographers* 95, no. 3 (2005), 542–565.

——. "The 'Commons' versus the 'Commodity': 'Alter'-Globalization, Anti-Privatization and the Human Right to Water in the Global South." *Antipode* 39, no. 3 (2007a): 430–455.

——. "Trickle Down? Private Sector Participation and the Pro-Poor Water Supply Debate in Jakarta, Indonesia." *Geoforum* 38, no. 5 (2007b): 855–868.

——. "The Ambiguity of Community: Debating Alternatives to Water Supply Privatization." *Water Alternatives* 1, no. 2 (2008): 236–252.

Bakker, K., and D. Hemson. "Privatizing Water: Hydropolitics in the New South Africa." *South African Journal of Geography* 82, no. 1 (2000): 3–12.

Bakker, K., M. Kooy, M. Shofiani, and E.J. Martjin. "Governance Failure: Rethinking the Institutional Dimensions of Urban Water Supply to Poor Households." *World Development* 36, no. 10 (2008): 1891–1915.

Balance, T., and S. Trémolet. *Private Sector Participation in Urban Water Supply in Sub-Saharan Africa*. Frankfurt: German Development Cooperation, 2005.

Baland, J., and J. Platteau. "Wealth Inequality and Efficiency in the Commons Part I: The Unregulated Case." *Oxford Economic Papers* 49, no. 4 (1997): 451–482.

——. "Wealth Inequality and Efficiency in the Commons Part II: The Regulated Case." *Oxford Economic Papers* 50, no. 1 (1998): 1–22.

Balanyá, B., B. Brennan, O. Hoedeman, S. Kishimoto, and P. Terhorst. *Reclaiming Public Water: Achievements, Struggles and Visions from around the World*. Amsterdam: Transnational Institute and Corporate Europe Observatory, 2005.

Banerjee, S., J. Oetzel, and R. Ranganathan. "Private Provision of Infrastructure in Emerging Markets: Do Institutions Matter?" *Development Policy Review* 24, no. 2 (2006): 175–202.

Bangura, Y., and G. Larbi, eds. *Public Sector Reform in Developing Countries*. London: Palgrave Macmillan, 2006.

Barkin, David, ed. *La gestión del Agua Urbana en México: Retos, debates y bienestar*. Guadalajara: Universidad de Guadalajara, 2006.

Barlow, M. *Blue Covenant: The Global Water Crisis and the Coming Battle for the Right to Water*. Toronto: McClelland and Stewart, 2007.

Barlow, M., and T. Clarke. *Blue Gold: The Fight to Stop the Corporate Theft of the World's Water*. New York: Stoddart, 2003.

Barnes, P. *Capitalism 3.0: Reclaiming the Commons*. San Francisco: Berrett-Koehler, 2006.

Barrett, D., and V. Jaichand. "The Right to Water, Privatised Water and Access to Justice: Tackling United Kingdom Water Companies' Practices in Developing Countries." *South African Journal on Human Rights* 23, no. 3 (2007): 543–562.

Bashford, A. *Imperial Hygiene: A Critical History of Colonialism, Nationalism, and Public Health.* London: Macmillan, 2004.

Batley, R. "Public-Private Relationships and Performance in Service Provision." *Urban Studies* 33, no. 4–5 (1996): 723–751.

Batley, R., and G. Larbi. "Capacity to Deliver? Management, Institutions, and Public Services in Developing Countries." In *Public Sector Reform in Developing Countries,* edited by Y. Bangura and G. Larbi, 99–130. London: Palgrave Macmillan, 2006.

Bauer, C. "Bringing Water Markets Down to Earth: The Political Economy of Water Rights in Chile, 1976–1995." *World Development* 25, no. 5 (1997): 639–656.

——. "Slippery Property Rights: Multiple Water Uses and the Neoliberal Model in Chile, 1981–1995." *Natural Resources Journal* 38 (1998): 109–156.

Bayliss, K. "Privatization and Poverty: The Distributional Impact of Utility Privatization." *Annals of Public & Cooperative Economics* 73, no. 4 (2002): 603–625.

Bayliss, K., and B. Fine. *Privatization and Alternative Public Sector Reform in Sub-Saharan Africa: Delivering on Electricity and Water.* London: Palgrave Macmillan, 2007.

Beck, U. *Risk Society: Towards a New Modernity.* London: Sage, 1992.

Bennett, V., S. Dávila-Poblete, and N. Rico, eds. *Opposing Currents: The Politics of Water and Gender in Latin America.* Pittsburgh: University of Pittsburgh Press, 2005.

Bennholdt-Thomsen, V., and M. Mies. *The Subsistence Perspective: Beyond the Globalized Economy.* London: Zed Books, 2000.

Berkes, F. *Common Property Resources: Ecology and Community-Based Sustainable Development.* London: Belhaven Press, 1989.

Bernstein, S. *The Compromise of Liberal Environmentalism.* New York: Columbia University Press, 2001.

Berry, B. "Clean Water for All: Equity-Based Urban Water Supply Alternatives for Indonesia's Cities." *Urban Geography* 3, no. 4 (1982): 281–299.

Bhatia, R., R. Cessti, and C. Van der Berg. "Water Demand Management and Pollution Control in the Jabotabek Region, Indonesia." Unpublished World Bank paper, 1994.

Birchall, J., ed. *The New Mutualism in Public Policy.* London: Routledge, 2001.

——. "Mutual, Non-Profit or Public Interest Company? An Evaluation of Options for the Ownership and Control of Water Utilities." *Annals of Public and Cooperative Economics* 73, no. 2 (2002): 181–213.

Birchall, J., and R. Simmons. "What Motivates Members to Participate in Co-operative and Mutual Businesses?" *Annals of Public and Cooperative Economics* 75, no. 3 (2004): 465–495.

Birdsall, N., and J. Nellis. "Winners and Losers: Assessing the Distributional Impacts of Privatization." *World Development* 31, no. 10 (2003): 1617–1633.

Biswas, A. "Dams: Cornucopia or Disaster?" *International Journal of Water Resources Development* 20, no. 1 (2004): 3–14.

——. "Integrated Water Resources Management: Is It Working?" *International Journal of Water Resources Development* 24, no. 1 (2008), 5–22.

Biwater. "Statement about Biwater Gauff Tanzania." Press release. December 1. Surrey, UK: Biwater, 2005.

Bjornlund, H. "Efficient Water Market Mechanisms to Cope with Water Scarcity." *International Journal of Water Resources Development* 19, no. 4 (2003): 553–567.

Black, M. *Learning What Works: A 20-Year Retrospective View on International Water and Sanitation Cooperation; 1978–1998.* Washington, DC: UNDP-World Bank Water and Sanitation Program, World Bank, 1998.

Blokland, M., O. Braadbaart, and K. Schwartz. *Private Business, Public Owners.* The Hague: VROM, 1999.

Bluemel, E. "The Implications of Formulating a Human Right to Water." *Ecology Law Quarterly* 31, no. 4 (2004): 957–1006.

Boardman, B. *Fuel Poverty: From Cold Homes to Affordable Warmth.* London: Belhaven, 1991.

Boardman, B., S. Bullock, and D. McLaren. *Equity and the Environment: Guidelines for Green and Socially Just Government.* London: Friends of the Earth, 1999.

Boelens, R. "The Rules of the Game and the Game of the Rules: Normalization and Resistance in Andean Water Control." PhD diss., Wageningen University, Netherlands, 2008.

——. "The Politics of Disciplining Water Rights." *Development and Change* 40, no. 2 (2009): 307–331.

Boelens, R., and H. de Vos. "Water Law and Indigenous Rights in the Andes." *Cultural Survival Quarterly* 29, no. 4 (2006): 18–21.

Boelens, R., and B. Doornbos. "The Battlefield of Water Rights: Rule Making amidst Conflicting Normative Frameworks in the Ecuadorian Highlands." *Human Organization* 60, no. 4 (2001): 343–355.

Boelens, R., and P. Gelles. "Cultural Politics, Communal Resistance and Identity in Andean Irrigation Development." *Bulletin of Latin American Research* 24, no. 3 (2005): 311–327.

Boelens, R., and M. Zwarteveen. "Prices and Politics in Andean Water Reforms." *Development and Change* 36, no. 4 (2005): 735–758.

Boland, A. "From Provision to Exchange: Legalizing the Market in China's Urban Water Supply." In *Everyday Modernity in China*, edited by M. Dong and J. Goldstein, 303–331. Seattle: University of Washington Press, 2006.

——. "The Trickle-Down Effect: Ideology and the Development of Premium Water Networks in China's Cities." *International Journal of Urban and Regional Research* 31, no. 1 (2007): 21–40.

Bond, P., ed. *Unsustainable South Africa: Environment, Development, and Social Protest.* London: Merlin Press, 2002.

Bond, P. *Against Global Apartheid: South Africa Meets the World Bank, the IMF, and International Finance.* London: Zed Books, 2003a.

——. "Uneven Health Outcomes and Political Resistance under Residual Neoliberalism in Africa." *International Journal of Health Services* 33, no. 2 (2003b): 607–630.

——. "Water Commodification and Decommodification Narratives: Pricing and Policy Debates from Johannesburg to Kyoto to Cancun and Back." *Capitalism Nature Socialism* 15, no. 1 (2004): 7–25.

——. "Affidavit in the Case of Mazibuko et al. versus the City of Johannesburg," edited by High Court of South Africa (Witwatersrand Local Division), 2007.

Bond, P., and J. Dugard. "Water, Human Rights and Social Conflict: South African Experiences." *Law, Social Justice & Global Development* 1 (2008): 1–21.

Bond, P., and P. McInnes. "Decommodifying Electricity in Post-Apartheid Johannesburg." In *Contesting Neoliberalism: Urban Frontiers,* edited by H. Leitner, J. Peck, and E. Sheppard, 157–178. New York: Guilford, 2007.

Bond, P., S. Nojiyeza, D. Khumalo, and B. Amisi. "The Neoliberal Loo." *ZMag,* February 2008. http://www.zmag.org/zspace/commentaries/3372.

BPD. "Jakarta, Indonesia: Water Supply Improvements to Marunda District." London: Building Partnerships for Development in Water and Sanitation, 2001. http://www.bpd-waterandsanitation.org/english/projects/jakarta.asp.

——. "Making Innovation Work through Partnerships in Water and Sanitation Projects." London: Building Partnerships for Development in Water and Sanitation, 2002. http://www.bpdwaterandsanitation.org/english/docs/innovation.pdf.

BPS. *Statistik kesejahteraan rakyat/Social Welfare Statistics.* Jakarta: Indonesian Bureau of Statistics, 2005.

Braadbaart, O., and F. Braadbaart. "Policing the Urban Pumping Race: Industrial Groundwater Overexploitation in Jakarta." *World Development* 25, no. 2 (1997): 199–210.

Brecher, T., T. Costello, and B. Smith. *Globalization from Below: The Power of Solidarity.* 2nd ed. Boston: South End Press, 2000.

Brennan, E.M., and H.W. Richardson. "Asian Megacity Characteristics, Problems and Policies." *International Regional Science Review* 12, no. 2 (1989): 117–129.

Briceno-Garmendia, C., A. Estache, and N. Shafik. "Infrastructure Services in Developing Countries: Access, Quality, Costs, and Policy Reform." World Bank Policy Research Paper WPS 3468, 2004.

Briscoe, J., and H.A. Garn. "Financing Water Supply and Sanitation under Agenda 21." *Natural Resources Forum* 19, no. 1 (1995): 59–70.

Brooks, D., O. Brandes, and S. Gurman, eds. *Making the Most of the Water We Have: The Soft Path Approach to Water Management.* London: Earthscan, 2009.

Brooks, D.B. "Human Rights to Water in North Africa and the Middle East: What Is New and What Is Not; What Is Important and What Is Not." In *Water as a Human Right for the Middle East and North Africa,* edited by A.K. Biswas, E. Rached, and C. Tortajada, 19–34. New York: Routledge, 2008.

Brown, C.J., and M. Purcell. "There's Nothing Inherent about Scale: Political Ecology, the Local Trap, and the Politics of Development in the Brazilian Amazon." *Geoforum* 36, no. 5 (2005): 607–624.

Budds, J. "Power, Nature and Neoliberalism: The Political Ecology of Water in Chile." *Singapore Journal of Tropical Geography* 25, no. 3 (2004): 322–342.

Budds, J., and G. McGranahan. "Are the Debates on Water Privatization Missing the Point? Experiences from Africa, Asia and Latin America." *Environment and Urbanization* 15, no. 2 (2003): 87–114.

Burki, S.J., and G. Perry. *Beyond the Washington Consensus: Institutions Matter.* Report 18482. Washington, DC: World Bank, 1998.

Bustamante, R., E. Peredo, and M. Udaeta. "Women in the 'Water War' in the Cochabamba Valleys." In *Opposing Currents: The Politics of Water and Gender in Latin America,* edited by V. Bennett, S. Dávila-Poblete, and M. Nieves Rico, 72–90. Pittsburgh: University of Pittsburgh Press, 2005.

Caffentzis, G. "A Tale of Two Conferences: Globalization, the Crisis of Neoliberalism and the Question of the Commons." Center for Global Justice, 2004.

Cairncross, S. "Water Supply and the Urban Poor." In *The Poor Die Young: Housing and Health in Third World Cities,* edited by S. Cairncross, J.E. Hardoy, and D. Satterthwaite, 109–126. London: Earthscan, 1990.

Calle, J., and M. Lohman. "The Triumph of the National Majorities and the Challenge of Change." In *Beyond the Market: The Future of Public Services,* edited by Daniel Chavez, 75–83. Amsterdam: Transnational Institute, 2006.

Cantin-Cumyn, M. "The Legal Status of Water in Quebec." *Quebec Studies, American Council for Quebec Studies* 42 (2006–2007): 7–15.

Castree, N., and B. Braun. *Social Nature: Theory, Practice and Politics.* Oxford: Wiley-Blackwell, 2001.

Castro, J.E. "Poverty and Citizenship: Sociological Perspectives on Water Services and Public-Private Participation." *Geoforum* 38, no. 5 (2007): 756–771.

——. "Neoliberal Water and Sanitation Policies as a Failed Development Strategy: Lessons from Developing Countries." *Progress in Development Studies* 8, no. 1 (2008): 63–83.

Castro, J.E., M. Kaika, and E. Swyngedouw. "Urban Water: A Political-Ecology Perspective." *Built Environment* 28, no. 2 (2002): 124–137.

Cesano, D., and J.E. Gustafsson. "Impact of Economic Globalisation on Water Resources: A Source of Technical, Social and Environmental Challenges for the Next Decade." *Water Policy* 2 (2000): 213–227.

CESCR. "Committee on Economic, Social and Cultural Rights Adopts General Comment on Right to Water." Press release. November 26. Geneva: United Nations Committee on Economic, Social and Cultural Rights, 2002a.

CESCR. *General Comment No. 15 (2002): The Right to Water (Articles 11 and 12 of the International Convention on Economic, Social and Cultural Rights).* Geneva: Economic and Social Council, 2002b.

Cessti, R. *Policies for Water Demand Management and Pollution Control in the Industrial and Household Sectors in the Jabotabek Region.* Jakarta: World Bank, 1993.

Chang, H. "The Economic Theory of the Developmental State." In *The Developmental State,* edited by M. Woo-Cummings, 182–199. Ithaca, NY: Cornell University Press, 1999.

Chang, H., and R. Rowthorn, eds. *The Role of the State in Economic Change.* Oxford: Oxford University Press, 1995.

Chant, C., ed. *Science, Technology and Everyday Life 1870–1950.* London: Routledge, 1989.

Chatterjee, P. *Politics of the Governed: Reflections on Popular Politics in Most of the World.* New York: Columbia University Press, 2004.

Checchi, D., M. Florio, and J. Carrera. *Privatization Discontent and Its Determinants: Evidence from Latin America.* Bonn: Institute for the Study of Labor, discussion paper series no. 1587, 2006.

Chifos, C. "Meeting Unmet Expectations: Initiatives in Urban Environmental Service Provision in Jakarta." In *Southeast Asian Urban Environments,* edited by C. Chifos and R. Yabes, 29–72. Tempe: Arizona State University Press, 2000.

Chifos, C., and S. Hendropranoto. "Thirty Years of Urban Infrastructure Development in Indonesia." In *Southeast Asian Urban Environments,* edited by C. Chifos and R. Yabes, 153–181. Tempe: Arizona State University Press, 2000.

Christoff, P. "Ecological Modernisation, Ecological Modernities." *Environmental Politics* 5, no. 3 (1996): 476–500.

Cipolla, C. *Miasmas and Disease: Public Health and the Environment in the Pre-Industrial Age.* New Haven, CT: Yale University Press, 1992.

City of Johannesburg. "Phiri Case." Press release. May 2008. http://www.joburg.org.za/content/view/2485/114/.

Clarke, G., K. Kosec, and S. Wallsten. "Has Private Participation in Water and Sewerage Improved Coverage? Empirical Evidence from Latin America." World Bank Policy Research Working Paper 3445. Washington, DC: World Bank, 2004.

Cleaver, F. "Moral Ecological Rationality, Institutions and the Management of Common Property Resources." *Development and Change* 31, no. 2 (2000): 361–383.

Cleaver, F., and A. Toner. "The Evolution of Community Water Governance in Uchira, Tanzania: The Implications for Equality of Access, Sustainability and Effectiveness." *Natural Resources Forum* 30, no. 3 (2006): 207–218.

Cochrane, A. "Community Politics and Democracy." In *New Forms of Democracy,* edited by D. Held and C. Pollitt, 51–77>. London: Sage, 1986.

COHRE, AAAS, SDC, and UN-HABITAT. *Manual on the Right to Water and Sanitation.* Geneva: Centre on Housing Rights and Evictions, 2008.

Collignon, B. "The Potential and the Limits of Private Water Providers—Independent Markets in Francophone Africa." Nairobi and Washington, DC: UNDP/WB Water and Sanitation Programme, 1998.

Collignon, B., and M. Vézina. "Independent Water and Sanitation Providers in African Cities: Full Report of a Ten-Country Study." Washington, DC: UNDP/WB Water and Sanitation Programme, April 2000.

Conca, K. *Governing Water: Contentious Transnational Politics and Global Institution Building.* Boston: MIT Press, 2006.

Cooke, B., and U. Kothari. *Participation: The New Tyranny?* London: Zed Books, 2001.

Co-operatives Secretariat. "Co-operatives in Canada 1999." Ottawa: Co-operatives Secretariat, Government of Canada, 2001.

Corbridge, S., G. Williams, M. Srivastava, and R. Véron. *Seeing the State: Governance and Governability in India.* Cambridge: Cambridge University Press, 2005.

Corry, D., W. Hatter, I. Parker, A. Randle, and G. Stoker. "Joining Up Local Democracy: Governance Systems for New Localism in New Local Government Network." London: New Local Government Network, 2004.

Cote, J. "A Conglomeration of... Often Conflicting Ideas: Resolving the 'Native Question' in Java and the Outer Islands in the Dutch East Indies, 1900–1925." *Itinerario* 3/4 (2003): 160–188.

Cour de Cassation (chamber criminelle), du 08 avril 1999, pourvoi n\o60 98–84539, Jean-Jacques Prompsy et autres.

Coutard, O. *The Governance of Large Technical Systems.* London: Routledge, 1999.

Cowan, S. "Competition in the Water Industry." *Oxford Review of Economic Policy* 13, no. 1 (1997): 83–92.

Cowherd, R. "Cultural Construction of Jakarta: Design, Planning and Development in Jabotabek, 1980–1997." PhD diss., Massachusetts Institute of Technology, 2002.

Crane, R. "Water Markets, Market Reform and the Urban Poor: Results from Jakarata, Indonesia." *World Development* 22, no. 1 (1994): 71–83.

Crane, R., and A. Daniere. "The Contribution of Environmental Amenities to Low-Income Housing: A Comparative Study of Bangkok and Jakarta." *Urban Studies* 34, no. 9 (1997): 1495–1512.

Cremers, L., M. Ooijevaar, and R. Boelens. "Institutional Reform in the Andean Irrigation Sector: Enabling Policies for Strengthening Local Rights and Water Management." *Natural Resources Forum* 29, no. 1 (2005): 37–50.

Crespo Flores, C., and R. Orellana Halkyer. *Conflictos ambientales, dos casos: Agua y territorio.* Cochabamba, Bolivia: CERES, 1999.

Cross, P., and A. Morel. "Pro-Poor Strategies for Urban Water Supply and Sanitation Services in Africa." *Water Science and Technology* 51, no. 8 (2005): 51–57.

Curry, J., and S. McGuire. *Community on Land: Community, Ecology, and the Public Interest.* Lanham, MD: Rowman and Littlefield, 2002.

Cutler, D., and G. Miller. "Water, Water Everywhere: Municipal Finance and Water Supply in American Cities." In *Corruption and Reform: Lessons from America's History,* edited by E. Glaeser and C. Goldin, 153–186. Chicago: University of Chicago Press, 2006.

Darrow, M. "Power, Capture, and Conflict: A Call for Human Rights Accountability in Development Cooperation." *Human Rights Quarterly* 27, no. 2 (2005): 471–538.

Davis, J. "Private-Sector Participation in the Water and Sanitation Sector." *Annual Review of Environment and Resources* 30 (2005): 145–183.

Davis, M. *Planet of Slums.* London: Verso, 2006.

del Moral, L., and D. Sauri. "Recent Developments in Spanish Water Policy: Alternatives and Conflicts at the End of the Hydraulic Era." *Geoforum* 32, no. 3 (2001): 351–362.

De Loe, R., S. Di Giantomasso, and R. Kreutzwiser. "Local Capacity for Groundwater Protection in Ontario." *Environmental Management* 29, no. 2 (2002): 217–233.

Dennis, M., and D. Stewart. "Justiciability of Economic, Social, and Cultural Rights: Should There Be an International Complaints Mechanism to Adjudicate the Rights to Food, Water, Housing, and Health?" *American Journal of International Law* 98, no. 3 (2004): 462–515.

Derman, B., and A. Hellum. "Livelihood Rights Perspective on Water Reform: Reflections on Rural Zimbabwe." *Land Use Policy* 24, no. 4 (2007): 664–673.

de Vos, H., R. Boelens, and R. Bustamante. "Formal Law and Local Water Control in the Andean Region: A Fiercely Contested Field." *International Journal of Water Resources Development* 22, no. 1 (2006): 37–48.

DFID. "Better Water Services in Developing Countries: Public-Private Partnership the Way Ahead." London: Department for International Development, 1998.

Dietz, T., N. Dolsak, E. Ostrom, and P. Stern. "The Drama of the Commons." In *The Drama of the Commons,* edited by E. Ostrom, T. Dietz, N. Dolsak, P. Stern, S. Sonich, and E. Weber, 3–35. Washington, DC: National Academy Press, 2002.

Dinar, A. "The Political Economy of Water Pricing Reforms." Washington, DC: World Bank; Oxford: Oxford University Press, 2000.

DKI. *Jakarta membangun* [Jakarta in Progress]. Government of Jakarta, 1972.

DKI Jakarta. "Jakarta's Kampung Improvement Programme." Jakarta: Directorate of Physical Development, Jakarta Capital City Government and KIP Implementation Unit, 1976.

Dracup, S. "Water Supply in Great Britain 1650–1950: A Brief History in Six Parts." *British Water Supply* (April 1973): 24–27.

Drozdz, J., J. Woodcock, R. Sukarma, S.B. Rasuanto, R. Barfield, and A. Purmono "Enabling Water Utilities to Serve the Urban Poor." Jakarta: World Bank, 2006.

D'Souza, R. "Water in British India: The Making of a 'Colonial Hydrology.'" *History Compass* 4, no. 4 (2006): 621–628.

DTI. "The UK Fuel Poverty Strategy." London: Department of Trade and Industry, 2001.

Dubash, N., M. Dupar, S. Kothari, and T. Lissu. "A Watershed in Global Governance? An Independent Assessment of the World Commission on Dams." Washington, DC: World Resources Institute and Lokayan, 2001.

Dubreuil, C. "The Right to Water: From Concept to Implementation." Marseilles: World Water Council, 2006a.

———. "Synthesis on the Right to Water at the 4th World Water Forum, Mexico." Marseilles: World Water Council, 2006b.

DWAF. "You and Your Water Rights: South African Law Review." Pretoria: Department of Water Affairs and Forestry, 1995.

———. "12 Successful Cost Recovery Case Studies for Water Services in South Africa." Pretoria: Department of Water Affairs and Forestry, 1998.

ECOSOC. "General Comment 15." Geneva: United Nations Committee on Economic, Social and Cultural Rights, 2002.

EDF. "Reforming the World Bank's Lending for Water: An NGO Critique of the World Bank's Draft Water Policy Paper." Washington, DC: Environmental Defense Fund, 1992.

Eggink, E.J. *Na 25 jaar: Beknopt gedenkschrift ter gelegenheid van het 25 jarig bestaan der Gemeente Batavia* [After 25 Years: A Concise Commemorative on the Occasion of 25 Years Existence of the Municipality of Batavia]. Batavia: Gemeente Batavia, 1930.

Ellul, J. *The Technological Society.* Translated by John Wilkinson. New York: Knopf, 1964.

Emel, J., and R. Roberts. "Institutional Form and Its Effect on Environmental Change: The Case of Groundwater in the Southern High Plains." *Annals of the Association of American Geographers* 85, no. 4 (1995): 664–683.

ESP. "Formative Research Progress: Preliminary Findings." Jakarta: USAID-Environmental Services Project, 2006.

Estache, A. "PPI Partnerships versus PPI Divorces in LDCs." World Bank Policy Research Working Paper 3470. Washington, DC: World Bank, 2005.

Estache, A., A. Gomez-Lobo, and D. Leipziger. "Utilities Privatization and the Poor: Lessons and Evidence from Latin America." World Development 29, no. 7 (2001): 1179–1198.

Estache, A., L. Guasch, and L. Trujillo. "Price Caps, Efficiency Payoffs, and Infrastructure Contract Renegotiations in Latin America." World Bank Policy Research Working Paper 3129. Washington, DC: World Bank, 2003.

Estache, A., and M. Rossi. "How Different Is the Efficiency of Public and Private Water Companies in Asia?" World Bank Economic Review 16, no. 1 (2002): 139–148.

European Parliament, Council. "Directive 2000/60/EC of the European Parliament and of the Council of 23 October 2000 Establishing a Framework for Community Action in the Field of Water Policy." Official Journal L 327 (2000): 1–73.

Evans, J. "What Is Local about Local Environmental Governance? Observations from the Local Biodiversity Action Planning Process." Area 36, no. 3 (2004): 270–279.

Evans, P. Embedded Autonomy: States and Industrial Transformation. Princeton, NJ: Princeton University Press, 1995.

———. "Introduction: Development Strategies across the Public-Private Divide." World Development 24, no. 6 (1996): 1033–1037.

Evans, P., and B. Appleton. "Community Management Today: The Role of Communities in the Management of Improved Water Supply Systems." The Hague, Netherlands: IRC International Water and Sanitation Centre, 1993.

Faruqui, N., A. Biswas, and J. Bino, eds. Water Management in Islam. Ottawa: International Development Research Centre, 2001.

Faust, D., and R. Nagar. "Third World NGOs and U.S. Academics: Dilemmas and Politics of Collaboration." Ethics, Place and Environment 6, no. 1 (2003): 73–78.

Feeny, D., S. Hanna, and A.F. McEvoy. "Questioning the Assumptions of the 'Tragedy of the Commons' Model of Fisheries." Land Economics 72, no. 2 (1996): 187–205.

Ferguson, J. The Anti-Politics Machine: "Development," Depoliticiziation and Bureaucratic Power in Lesotho. Minneapolis: University of Minnesota Press, 1994.

Fine, B. Social Capital versus Social Theory: Political Economy and Social Science at the Turn of the Millennium. London: Routledge, 2001.

Finnegan, W. "Leasing the Rain." New Yorker, April 8, 2002.

Firman, T. "The Restructuring of Jakarta Metropolitan Area: A 'Global City' in Asia." Cities 15, no. 4 (1998): 229–243.

———. "Indonesian Cities Under the 'Krismon': A Great Urban Crisis in Southeast Asia." Cities 16, no. 2 (1999): 69–82.

———. "New Town Development in Jakarta Metropolitan Region (JMR): A Perspective of Spatial Segregation." Habitat International 28, no. 3 (2004): 349–368.

Firman, T., and I. Dharmapatni. "The Challenges to Sustainable Development in Jakarta Metropolitan Region." Habitat International 18, no. 3 (1994): 79–94.

Fischer, L. The Story of Indonesia. New York: Harper Brothers, 1959.

Fischhendler, I., and E. Feitelson. "The Formation and Viability of a Non-Basin Water Management: The U.S.-Canada Case." *Geoforum* 36, no. 6 (2005): 792–804.

Fleishacker, S. *A Short History of Distributive Justice.* Cambridge, MA: Harvard University Press, 2004.

Ford, L. "A Model of Indonesian City Structure." *Geographical Review* 83, no. 4 (1993): 374–396.

FORKAMI. "Survey Report: Social Assessment for the Jakarta Output Based Aid Project." Jakarta: Forum Komunikasi Kualitas Air Minum Indonesia, 2006.

——. Personal communication, January 10, 2007.

Foucault, M. "Governmentality." In *The Foucault Effect: Studies in Governmentality,* edited by Graham Burchell, Colin Gordon, and Peter Miller, 87–104. Chicago: University of Chicago Press, 1991.

Franceys, R. "Water Resources Occasional Papers: Private Sector Participation in the Water and Sanitation Sector." Loughborough, UK: Department for International Development (DFiD), 1997.

——. "Water and Public Private Partnerships." *Global Water Report* 92 (2000): 16–17.

——. "Customer Committees, Economic Regulation and the Water Framework Directive." *Water Science and Technology: Water Supply* 6, no. 5 (2006): 9–15.

——. "GATS, 'Privatisation' and Institutional Development for Urban Water Provision: Future Postponed." *Progress in Development Studies* 8, no. 1 (2008): 45–58.

Franceys, R., and E. Gerlach, eds. *Regulating Water and Sanitation for the Poor: Economic Regulation for Public and Private Partnerships.* London: Earthscan, 2008.

Gale, F., and M. M'Gonigle, eds. *Nature, Production and Power: Towards an Ecological Political Economy.* Cheltenham, UK: Edward Elgar, 2000.

Galiani, S., P. Gertier, and E. Schargrodsky. *Water for Life: The Impact of the Privatization of Water Services on Chile Mortality.* Stanford: Center for Research on Economic Development and Policy Reform, 2002. Republished in *Journal of Political Economy* 113, no. 1 (2005): 83.

Galvin, M., and A. Habib. "The Politics of Decentralisation and Donor Funding in South Africa's Rural Water Sector." *Journal of Southern African Studies* 29, no. 4 (2003): 865–884.

Gandy, M. "The Paris Sewers and the Rationalization of Urban Space." *Transactions of the Institute of British Geographers* 24, no. 1 (1999): 23–44.

——. *Concrete and Clay: Reworking Nature in New York City.* Cambridge, MA: MIT Press, 2002.

——. "Rethinking Urban Metabolism: Water, Space and the Modern City." *City* 8, no. 3 (2004): 363–379.

——. "The Bacteriological City and Its Discontents." *Historical Geography* 34 (2006): 14–25.

Garande, T., and S. Dagg. "Public Participation and Effective Water Governance at the Local Level: A Case Study from a Small Under-Developed Area in Chile." *Environment, Development and Sustainability* 7, no. 4 (2005): 417–431.

García, A., F. García, and L. Quiton. *La "guerra del agua" Abril de 2000: La crisis de la política en Bolivia.* La Paz: Programa de Investigación Estratégica en Bolivia, 2003.

Gardner, G., and T. Prugh. "Seeding the Sustainable Economy." In *State of the World Report 2008: Innovations for a Sustainable Economy,* 3–17. Washington, DC: Worldwatch Institute, 2008.

Gassner, K., A. Popov, and N. Pushak. *Does Private Sector Participation Improve Performance in Electricity and Water Distribution?* PPIAF Trends and Policy Options, 6. Washington, DC: International Bank for Reconstruction and Development/World Bank, 2009. http://www.ppiaf.org/documents/trends_and_policy/PSP_water_electricity.pdf.

Gaye, M., and F. Diallo. "Community Participation in the Management of the Urban Environment in Rufisque (Senegal)." *Environment and Urbanization* 9, no. 1 (1997): 9–30.

Gemeente Batavia. *Batavia als handels-, industrie-, en woonstad* [Batavia as a Commercial, Industrial and Residential Centre]. Batavia and Amsterdam: Kolff, 1937.

George, S. *A Fate Worse Than Debt.* London: Penguin, 1988.

———. *The Debt Boomerang.* London: Pluto, 1992.

Gerschenkron, A. *Economic Backwardness in Historical Perspective.* Cambridge, MA: Harvard University Press, 1962.

Ghosh, G., and A. Nigam. "Comments on 'Financing Water Supply and Sanitation under Agenda 21' by John Briscoe and Mike Garn." *Natural Resource Forum* 19, no. 2 (1995): 161–165.

Giansante, C., L. Babiano, and L. del Moral Ituarte. "L'évolution des modalités d'allocation des ressources en eau en Espagne." *Revue d'economie méridionale* 48, no. 191 (2000): 235–247.

Giarracca, N., and N. del Pozo. "To Make Waves: Water and Privatization in Tucumán, Argentina." In *Opposing Currents: The Politics of Water and Gender in Latin America,* edited by V. Bennett, S. Dávila-Poblete, and M. Nieves Rico, 91–106. Pittsburg: University of Pittsburgh Press, 2005.

Gibbins, R. "Local Governance and Federal Political Systems." *International Social Science Journal* 53, no. 167 (2001): 163–170.

Gibson-Graham, J.K. *A Postcapitalist Politics.* Minneapolis: University of Minnesota Press, 2006.

Gilbert, A., and J. Gugler. *Cities, Poverty and Development: Urbanisation in the Third World.* 2nd ed. Oxford: Oxford University Press, 1992.

Gilbert, A., and D. James. "Water Pollution in Jakarta Bay." In *The Application of Economic Techniques in Environmental Impact Assessment,* edited by D. James, 111–142. Dordrecht: Kluwer, 1994.

Glassman, J. "Primitive Accumulation, Accumulation by Dispossession, Accumulation by 'Extra-Economic' Means." *Progress in Human Geography* 30, no. 5 (2006): 608–625.

Gleeson, B. "Community Care and Disability: The Limits to Justice." *Progress in Human Geography* 21, no. 2 (1997): 199–224.

Gleick, P. "Water and Conflict: Fresh Water Resources and International Security." *International Security* 18, no. 1 (1993): 79–112.

———. "Basic Water Requirements for Human Activities: Meeting Basic Needs." *Water International* 21 (1996): 83–92.

——. "The Human Right to Water." *Water Policy* 1 (1998): 487–503.

——. "The Changing Water Paradigm: A Look at Twenty-First Century Water Resources Development." *Water International* 25, no. 1 (2000): 127–138.

Goldman, M. *Imperial Nature: The World Bank and Struggles for Social Justice in the Age of Globalization.* New Haven, CT: Yale University Press, 2005.

Goldsmith, E., and N. Hilyard. *Social and Environmental Effects of Large Dams.* Cornwall: Wadebridge Ecological Centre, 1984.

Gopal, S. *Jawaharlal Nehru: An Anthology.* Delhi: Oxford University Press, 1980.

Gore, C. "The Rise and Fall of the Washington Consensus as a Paradigm for Developing Countries." *World Development* 28, no. 5 (2000): 789–804.

Goubert, J.P. *The Conquest of Water: The Advent of Health in the Industrial Age.* London: Polity, 1989.

Graf, W. "Damage Control: Restoring the Physical Integrity of America's Rivers." *Annals of the Association of American Geographers* 91, no. 1 (2001): 1–27.

Graham, S. "The Spectre of the Splintering Metropolis." *Cities* 18, no. 6 (2001): 365–368.

——. "On Technology, Infrastructure, and the Contemporary Urban Condition: A Response to Coutard." *International Journal of Urban and Regional Research* 26, no. 1 (2002): 175–182.

Graham, S., and S. Marvin. *Splintering Urbanism: Networked Infrastructures, Technological Mobilities and the Urban Condition.* London: Routledge, 2001.

Gready, P., and J. Ensor, eds. *Reinventing Development? Translating Rights-Based Approaches from Theory into Practice.* London: Zed Books, 2005.

Gregg, P. *A Social and Economic History of Britain: 1760–1972.* London: Harrap, 1950.

Groom, E., J. Halpern, and D. Ehrhardt. "Explanatory Notes on Key Topics in the Regulation of Water and Sanitation Services." World Bank Water Supply and Sanitation Sector Board Discussion Paper Series, Paper no. 6. Washington, DC: World Bank, 2006.

Grossman, E. *Watershed: The Undamming of America.* New York: Counterpoint, 2002.

Grusky, S. "Privatization Tidal Wave: IMF/Bank Water Policies and the Price Paid by the Poor." *Multinational Monitor* 22, no. 9 (2001). http://www.multinationalmonitor.org/mm2001/092001/grusky.html.

Guha, R. *Environmentalism: A Global History.* New York: Longman, 2000.

Gulyani, S., D. Talukdar, and R.M. Kariuki. "Universal (Non)Service? Water Markets, Household Demand and the Poor in Urban Kenya." *Urban Studies* 42, no. 8 (2005): 1247–1274.

Gupta, S., C. Schiller, H. Ma, and E. Tiongson. "Privatization, Labor and Social Safety Nets." *Journal of Economic Surveys* 15, no. 5 (2001): 647–670.

Gutierrez, E. *Washington D.C.'s Continuous Internal Improvement Alternative: An Initial Inquiry on PSP in Water and Sanitation in the US.* London: WaterAid, 2002.

Haarmeyer, D., and A. Mody. "Private Capital in Water and Sanitation." *Finance and Development* (March 1997): 34–37.

——. "World Water Privatization: Managing Risks in Water and Sanitation." London: Financial Times Energy, a division of Financial Times Business Limited, 1998.

Habbakuk, H. "Review of Rostow's Stages of Economic Growth." *Economic Journal* 71, no. 283 (1961): 601–604.

Habermas, J. *The Structural Transformation of the Public Sphere: An Inquiry into a Category of Bourgeois Society.* 1962. Translated by T. Burger. Cambridge, MA: MIT Press, 1989.

Haddad, B.M., ed. *River of Gold: Designing Markets to Allocate Water in California.* Washington, DC: Island Press, 2000.

Hajer, M.A. *The Politics of Environmental Discourse: Ecological Modernisation and the Policy Process.* Oxford: Clarendon Press, 1995.

Hall, D., and R. de la Motte. *Dogmatic Development: Privatisation and Conditionalities in Six Countries; PSIRU Report for War on Want.* London: PSIRU, 2004.

Hall, D., and E. Lobina. *Pipe Dreams: The Failure of the Private Sector to Invest in Water Services in Developing Countries.* London: World Development Movement and Public Services International Research Unit, 2006a.

———. *Water as a Public Service.* London: Public Services International Research Unit, University of Greenwich, 2006b.

———. "Profitability and the Poor: Corporate Strategies, Innovation and Sustainability." *Geoforum* 38, no. 5 (2007): 772–785.

Hall, D., E. Lobina, and R. de la Motte. "Public Resistance to Privatisation in Water and Energy." *Development in Practice* 15, no. 3–4 (2005): 286–301.

Hall, D., E. Lobina, O. Viero, and H. Maltz. *Water in Porto Alegre, Brazil: Accountable, Effective, Sustainable, and Democratic.* Paper prepared for the Johannesburg World Summit on Sustainable Development. London and Porto Alegre: Public Services International Research Unit, University of Greenwich and DMAE, 2002.

Hammer, L. "Indigenous Peoples as a Catalyst for Applying the Human Right to Water." *International Journal on Minority and Group Rights* 10 (2004): 131–161.

Hanna, W. "Bung Karno's Indonesia: The Impact of Djakarta." Southeast Asia Series. New York: American Universities Field Staff, 1959.

Hardin, G. "The Tragedy of the Commons." *Science* 162, no. 3859 (1968): 1243–1248.

Harsanto, D. "Pollution in City Rivers Worse: BPLHD." *Jakarta Post,* December 30, 2005.

Harsanto, D., and L. Wahyudi. "Water Is Going down the Drain: Union Leader." *Jakarta Post,* September 6, 2002.

Harsoso, A. "When Water and Political Power Intersect: A Journalist Probes the Story of Water Privatization in Jakarta Indonesia." *Nieman Reports* 59, no. 1 (2005): 45–47.

Harvey, D. *Social Justice and the City.* Baltimore: Johns Hopkins University Press, 1973.

———. *Justice, Nature and the Geography of Difference.* Oxford: Blackwell, 1996.

———. *The New Imperialism.* Oxford: Oxford University Press, 2003a.

———. "The Right to the City." *International Journal of Urban and Regional Research* 27, no. 4 (2003b): 939–941.

———. *A Short History of Neoliberalism.* Oxford: Oxford University Press, 2005.

Haṣṣan, J. *A History of Water in Modern England and Wales.* Manchester, UK: Manchester University Press, 1998.

Hausman, D.M., and M.S. McPherson. *Economic Analysis, Moral Philosophy, and Public Policy.* 2nd ed. New York: Cambridge University Press, 2006.

Hawkin, P., A. Lovins, and H. Lovins. *Natural Capital: Creating the Next Industrial Revolution.* Boston: Little, Brown, 1999.

Head, J.W. "Evolution of the Governing Law for Loan Agreements of the World Bank and Other Multilateral Development Banks." *American Journal of International Law* 90, no. 2 (1996): 214–234.

Heetjans, H. "Ingezonden: Watertarieven." *De waterstaats ingenieur* 3 (1923): 91–92.

Heikkila, T. "Institutional Boundaries and Common-Pool Resource Management: A Comparative Analysis of Water Management Programs in California." *Journal of Policy Analysis and Management* 23, no. 1 (2004): 97–117.

Heller, P. "Moving the State: The Politics of Democratic Decentralization in Kerala, South Africa, and Porto Alegre." *Politics and Society* 29, no. 1 (2001): 131–163.

Hemson, D. "Beating the Backlog: Meeting Targets and Providing Free Basic Services." Position paper for the National Treasury. Pretoria, 2004.

Hemson, D., and B. Dube. "Water Services and Public Health: The 2000–01 Cholera Outbreak in Kwazulu Natal, South Africa." Paper presented at the eighth World Congress on Environmental Health, February 22–27. Durban, South Africa: Human Sciences Research Council Report, 2004.

Heynen, N., J. McCarthy, W.S. Prudham, and P. Robbins, eds. *Neoliberal Environments: False Promises and Unnatural Consequences.* New York: Routledge, 2007.

Himley, M. "Geographies of Environmental Governance: The Nexus of Nature and Neoliberalism." *Geography Compass* 2, no. 2 (2008): 433–451.

Hodge, G. *Privatization: An International Review of Performance.* Boulder, CO: Westview Press, 2000.

Hoedeman, O. "Public Water for All: The Role of Public-Public Partnerships." Amsterdam: Transnational Institute and Corporate Europe Observatory, 2006.

Holland, A. *The Water Business: Corporations versus People.* London: Zed Books, 2005.

Holloway, M. "Nurturing Nature." *Scientific American* 270, no. 4 (1994): 98–108.

Holmes, P.R. "Effective Organizations for Water Management." *International Journal of Water Resources Development* 16, no. 1 (2000): 57–71.

Huffaker, R., and N. Whittlesey. "A Theoretical Analysis of Economic Incentive Policies Encouraging Agricultural Water Conservation." *International Journal of Water Resources Development* 19, no. 1 (2003): 37–53.

Hughes, D.M. "Whites and Water: How Euro-Africans Made Nature at Kariba Dam." *Journal of Southern African Studies* 32, no. 4 (2006): 823–838.

Hukka, J.J., and T.S. Katko. "Refuting the Paradigm of Water Services Privatization." *Natural Resource Forum* 27, no. 2 (2003): 142–155.

Hunt, L.C., and E. Lynk "Privatisation and Efficiency in the UK Water Industry: An Empirical Analysis." *Oxford Bulletin of Economics and Statistics* 57, no. 3 (1995): 371–388.

Ibrahim, S. "From Individual to Collective Capabilities: The Capability Approach as a Conceptual Framework for Self-Help." *Journal of Human Development* 7, no. 3 (2006): 397–416.

IBRD. "Water Supply and Sewerage, Sector Program Paper R7184," 1971. Cited in Kapur, Lewis, and Webb, *The World Bank: Its First Half Century*. Vol. 1, *History*, 21 n. 217.

ICSID. *Biwater Gauff (Tanzania) Limited v. United Republic of Tanzania (ICSID Case No. ARB/05/22)*. Washington, DC: International Centre for Settlement of Investment Disputes, 2008.

Ignatieff, M. *Human Rights as Politics and Idolatry*. Princeton, NJ: Princeton University Press, 2003.

Illich, I. *H2O and the Waters of Forgetfulness*. Berkeley: Heydey Books, 1985.

IMF. "Public-Private Partnerships." Washington, DC: International Monetary Fund (Fiscal Affairs Department), 2004.

Izaguirre, K. "Private Infrastructure: Emerging Market Sponsors Dominate Private Flows. Public Policy for the Private Sector." Washington, DC: World Bank, 2005.

Jaglin, S. "The Right to Water versus Cost Recovery: Participation, Urban Water Supply and the Poor in Sub-Saharan Africa." *Environment and Urbanization* 14, no. 1 (2002): 231–245.

Jakarta Water Supply Regulatory Body (JWSRB). *Automatic Tariff Adjustments*. Jakarta: Jakarta Water Supply Regulatory Board, 2004a.

——. "Policy Analysis: Implementation of Automatic Tariff Adjustment for DKI Jakarta Drinking Water." Jakarta: Regulatory Board for DKI Jakarta Water Supply, 2004b.

——. "Urban Water in Jakarta." Presentation at the Urban Water and Citizenship Workshop on May 11, 2005. Jakarta: Jakarta Water Supply Regulatory Body (Badan Regulator), 2005a.

——. "Current Status Restated Cooperation Agreement (RCA)." Jakarta: Water Service Provision in DKI, 2005b.

——. "Jakarta Water Supply Report 1998–April 2006." Jakarta: Jakarta Water Supply Regulatory Body, 2006a.

——. "Regulatory Approach to the Jakarta Water Supply Concession Contracts." *International Journal of Water Resources Development* 22, no. 2 (2006b): 255–276.

——. *The First Ten Years of Implementation of the Jakarta Water Supply 25-Year Concession Agreement (1998–2008)*. Jakarta: Jakarta Water Supply Regulatory Body, 2008.

Jasanoff, S., ed. *States of Knowledge: The Co-production of Science and Social Order*. New York: Routledge, 2004.

Jellinek, L. *The Wheel of Fortune: The History of a Poor Community in Jakarta*. Honolulu: University of Hawaii Press, 1991.

——. "Big Projects, Little People: The Battle for Kemayoran." *Inside Indonesia* 50 (April–June 1997). http://insideindonesia.org/content/view/875/29/.

Jensen, O., and F. Blanc-Brude. "The Handshake: Why Do Governments and Firms Sign Private Sector Participation Deals? Evidence from the Water and Sanitation Sector in Developing Countries." World Bank Policy Research Working Paper 3937. Washington, DC: World Bank, 2006.

Jessop, B. "The Regulation Approach, Governance and Post-Fordism: Alternative Perspectives on Economic and Political Change?" *Economy and Society* 24, no. 3 (1995): 307–333.

Jessop, R.D. "Governance Failure." In *Encyclopedia of Governance,* edited by M. Bevir, 381–382. London: Sage, 2006.

JICA. "Study: Jakarta Water Supply Development Project." Vol. 2. Main report. Jakarta: Nihon Suido Consultants and Nippon Koei, 1997.

Johnson, C., and D. Start. "Rights, Claims and Capture: Understanding the Politics of Pro-Poor Policy." Working paper 145. London: Overseas Development Institute, 2001.

Johnston, J., M. Gismondi, and J. Goodman. *Nature's Revenge: Reclaiming Sustainability in an Age of Corporate Globalization.* Toronto: Broadview, 2006.

Johnstone, N., and L. Wood, eds. *Private Firms and Public Water: Realising Social and Environmental Objectives in Developing Countries.* London: International Institute for Environment and Development, 2001.

Juuti, P., T. Katko, and R. Rajala. "For the Quality of Life—Evolution of and Lessons Learnt from Water and Sanitation Services in Porvoo, Finland, 1900–2000." *Natural Resources Forum* 29, no. 2 (2005): 109–119.

Kaika, M. *City of Flows: Modernity, Nature and the City.* London: Routledge, 2005.

Kaika, M., and E. Swyngedouw. "Fetishizing the Modern City: The Phantasmagoria of Urban Technological Networks." *International Journal of Urban and Regional Research* 24, no. 1 (2000): 120–138.

Kappagoda, N. *The Multilateral Development Banks.* Vol. 2, *The Asian Development Bank.* Boulder, CO: Lynne Rienner, 1996.

Kapur, D., J. Lewis, and R. Webb. *The World Bank: Its First Half Century.* Vol. 1, *History.* Washington, DC: Brookings Institution, 1997.

Kariuki, M., and J. Schwartz. "Small-Scale Private Service Providers of Water Supply and Electricity: A Review of Incidence, Structure, Pricing and Operating Characteristics." World Bank Policy Research Working Paper WPS3727. Washington, DC: World Bank, 2005.

Karsten, T. "Town Development in the Indies." In *The Indonesian Town: Studies in Urban Sociology,* edited by W.F. Wertheim, 3–77. The Hague: W. van Hoeve, 1958.

Katko, T.S. "Evolution of Consumer-Managed Water Cooperatives in Finland, with Implications for Developing Countries." *Water International* 17, no. 1 (1992): 12–20.

———. *Water! Evolution of Water Supply and Sanitation in Finland from the Mid-1800s to 2000.* Tampere: Finnish Water and Waste Water Works Association, 2000.

Katz, C. "Whose Nature, Whose Culture? Private Productions of Space and the 'Preservation' of Nature." In *Remaking Reality,* edited by B. Braun and N. Castree, 46–63. London: Routledge, 1998.

Kaul, I., and P. Concei<c &cedilla>ão. *The New Public Finance: Responding to Global Challenges.* Oxford: Oxford University Press, 2006.

Kay, J. "Regulating Private Utilities: The Customer Corporation." *Journal of Co-operative Studies* 29, no. 2 (1996): 28–46.

Keil, R., and G. Desfor. *Nature and the City: Making Environmental Policy in Toronto and Los Angeles.* Tucson: University of Arizona Press, 2004.

Kemper, K., A. Dinar, and W. Blomquist, eds. *Institutional and Policy Analysis of River Basin Management Decentralization: The Principle of Managing Water Resources*

at the Lowest Appropriate Level; When and Why Does It (Not) Work in Practice?
Washington, DC: World Bank, 2005.

Kerf, M., and A. Izaguirre. "Revival of Private Participation in Developing Country Infrastructure: A Look at Recent Trends and Policy Implications." *Gridlines* 16. Washington, DC: World Bank, Public Private Infrastructure Advisory Facility, 2007.

Kessides, I. *Reforming Infrastructure: Privatization, Regulation, and Competition.* World Bank Policy Research Report. Washington, DC: The International Bank for Reconstruction and Development, 2004.

——. "Infrastructure Privatization and Regulation: Promises and Perils." *World Bank Research Observer* 20, no. 1 (2005): 81–108.

Khagram, S. *Dams and Development: Transnational Struggles for Water and Power.* Ithaca, NY: Cornell University Press, 2004.

Kijne, J. "Lessons Learned from the Change from Supply to Demand Water Management in Irrigated Agriculture: A Case Study from Pakistan." *Water Policy* 3, no. 2 (2001): 109–123.

Kinnersley, D. *Troubled Water: Rivers, Politics and Pollution.* London: Hilary Shipman, 1988.

KIP. *Jakarta's Kampung Improvement Programme.* Jakarta: Directorate of Physical Development, Jakarta Capital City Government and KIP Implementation Unit, 1976.

Kirkpatrick, C., and P. Cook. *Privatization in Less Developed Countries.* London: Palgrave Macmillan, 1988.

Kirkpatrick, C., D. Parker, and Y. Zhang. "An Empirical Analysis of State and Private-Sector Provision of Water Services in Africa." *World Bank Economic Review* 20, no. 1 (2006): 143–163.

Kjellén, M. *From Public Pipes to Private Hands: Water Access and Distribution in Dar es Salaam, Tanzania.* Stockholm Studies in Human Geography. Stockholm: University of Stockholm, 2006.

Kliot, N., D. Shmueli, and U. Shamir. "Institutions for Management of Transboundary Water Resources: Their Nature, Characteristics and Shortcomings." *Water Policy* 3, no. 3 (2001): 229–255.

Kloezen, W. "Water Markets between Mexican Water User Associations." *Water Policy* 1 (1998): 437–455.

Kneen, B. *The Tyranny of Rights.* Ottawa: Ram's Horn Publications, 2009.

Komives, K., V. Fostser, J. Halpern, and Q. Wodon. "Water, Electricity, and the Poor: Who Benefits from Utility Subsidies?" Washington, DC: World Bank, 2005.

Kooy, M., and K. Bakker. "Splintered Networks: Urban Water Governance in Jakarta." *Geoforum* 39, no. 6 (2008a): 1843–1858.

——. "Technologies of Government: Constituting Subjectivities, Spaces, and Infrastructures in Colonial and Contemporary Jakarta." *International Journal of Urban and Regional Research* 32, no. 2 (2008b): 375–391.

Kreimer, A., R. Gilbert, C. Volonte, and G. Brown. "Enhancing the Quality of Life in Urban Indonesia: The Legacy of the Kampung Improvement Program." Rep.14747-IND. Washington, DC: Operations Evaluation Department, World Bank, 1995.

Kumar, M.D., and O.P. Singh. "Market Instruments for Demand Management in the Face of Scarcity and Overuse of Water in Gujarat, Western India." *Water Policy* 3, no. 5 (2001): 387–403.

Kusno, A. "Modern Beacon and the Traditional Polity: Jakarta in the Time of Sukarno." *Journal of Southeast Asian Architecture* 1 (1997): 387–403.

——. *Behind the Postcolonial: Architecture, Urban Space and Political Cultures in Indonesia.* London: Routledge, 2000.

Kuznets, S. *Modern Economic Growth: Rate, Structure, and Spread.* New Haven, CT: Yale University Press, 1966.

Kyessi, A.G. "Community-Based Urban Water Management in Fringe Neighbourhoods: The Case of Dar es Salaam, Tanzania." *Habitat International* 29, no. 1 (2005): 1–25.

Kymlicka, W. *Multicultural Citizenship: A Liberal Theory of Minority Rights.* New York: Oxford University Press, 1995.

Lacey, M. "A Changing Mind-Set." *American Water Works Association* 96, no. 4 (2004): 2.

Landry, C. "Market Transfers of Water for Environmental Protection in the Western United States." *Water Policy* 1, no. 5 (1998): 457–469.

Langa, D. "Judgment: Case No Cct 8/97." Edited by the Constitutional Court of South Africa, 1997.

Langford, M. "The United Nations Concept of Water as a Human Right: A New Paradigm for Old Problems?" *International Journal of Water Resources Development* 21, no. 2 (2005): 273–282.

——. "Ambition That Overleaps Itself? A Response to Stephen Tully's Critique of the General Comment on the Right to Water." *Netherlands Quarterly of Human Rights* 24, no. 3 (2006), 433–459.

Latour, B. *We Have Never Been Modern.* Cambridge, MA: Harvard University Press, 1993.

Laurie, N., and S. Marvin. "Globalisation, Neoliberalism and Negotiated Development in the Andes: Water Projects and Regional Identity in Cochabamba." *Environment and Planning A* 31, no. 8 (1999): 1401–1415.

Laxer, G., and D. Soron. *Not for Sale: Decommodifying Public Life.* Toronto: Broadview, 2006.

Leclerc, J. "Mirrors and the Lighthouse: A Search for Meaning in the Monuments and Great Works of Sukarno's Jakarta, 1960–65." In *Urban Symbolism,* edited by P. Nas, 38–58. Leiden: E.J. Brill, 1993.

Lee, T. "Alternatives for Private Participation in the Provision of Water Services." *Natural Resources Forum* 20, no. 4 (1996): 333–341.

Leitmann, J. "Urban Environmental Profile: A Global Synthesis of Seven Urban Environmental Profiles." *Cities* 12, no. 1 (1995): 23–39.

Lemos, M., and J. De Oliveira. "Water Reform across the State/Society Divide: The Case of Ceara, Brazil." *International Journal of Water Resources Development* 21, no. 1 (2005): 133–147.

Lenton, R.L., ed. *Health, Dignity, and Development: What Will It Take?* UN Millennium Project Task Force on Water and Sanitation. London: Earthscan, 2005.

Lewin, W., managing director, Ord Minnett Asset Management & Products. Speech at the Credit Suisse PL100 World Water Launch, 2007. http://www.credit-suisse.com/upload/news-live/000000021019.pdf.

Lewis, D., and D. Mosse. *Development Brokers and Translators: The Ethnography of Aid and Agencies*. Bloomfield, CT: Kumarian, 2006.

Li, T. "Compromising Power: Development, Culture and Rule in Indonesia." *Cultural Anthropology* 14, no. 3 (1999): 295–322.

——. *The Will to Improve: Governmentality, Development and the Practice of Politics*. Durham, NC and London: Duke University Press, 2007.

Lo, F.C., and Y.M. Yeung. "Global Restructuring and Emerging Urban Corridors in Pacific Asia." In *Emerging World Cities in Pacific Asia*, edited by Y.M. Yeung and F.C. Lo, 17–47. Tokyo: United Nations University Press, 1996.

Lobina, E., and D. Hall. "Public Sector Alternatives to Water Supply and Sewerage Privatization: Case Studies." *International Journal of Water Resources Development* 16, no. 1 (2000): 35–55.

——. "The Comparative Advantage of the Public Sector in the Development of Urban Water Supply." *Progress in Development Studies* 8, no. 1 (2008): 85–101.

Loftus, A., and D. McDonald. "Of Liquid Dreams: A Political Ecology of Water Privatization in Buenos Aires." *Environment and Urbanization* 13, no. 2 (2001): 179–199.

Lopez-Gunn, E. "The Role of Collective Action in Water Governance: A Comparative Study of Groundwater User Associations in La Mancha Aquifers in Spain." *Water International* 28, no. 3 (2003): 367–378.

Lorrain, D. "The Socio-Economics of Water Services: The Invisible Factors." Introduction to *Urban Water Management: French Experience around the World*, edited by D. Lorrain, 1–30. Levallois Perret, France: Hydrocom, 1997.

Lovei, L., and D. Whittington. "Rent-Extracting Behaviour by Multiple Agents in the Provision of Municipal Water Supply: A Study of Jakarta, Indonesia." *Water Resources Research* 29, no. 7 (1993): 1965–1974.

Lubis, D. "Turning the Bathroom into a Relaxing Retreat." *Jakarta Post*, April 18, 2004.

MacDonald, G. "Indonesia Medan Merdeka: National Identity and the Built Environment." *Antipode* 27, no. 3 (1995): 270–293.

Mansfield, B., ed. *Privatization: Property and the Remaking of Nature-Society Relations*. Malden, MA: Blackwell, 2008.

Marin, P. *Public-Private Partnerships for Urban Water Utilities: A Review of Experiences in Developing Countries*. Washington, DC: World Bank, 2009.

Marin, P., and A. Izaguirre. "Private Participation in Water: Toward a New Generation of Projects?" *Gridlines* 14. Washington, DC: World Bank, Public Private Infrastructure Advisory Facility, 2006.

Maronier, V. "De drinkwatervoorziening van Batavia" [The Drinking Water Network of Batavia]. *De waterstaats-ingenieur* 8 (1929): 223–239.

Marshall, G. "Farmers Cooperating in the Commons? A Study of Collective Action in Salinity Management." *Ecological Economics* 51, no. 3–4 (2004): 271–286.

Martijn, E. "Hydraulic Histories of Jakarta, Indonesia (1949–1997): Development of Piped Drinking Water Provision." Unpublished research report, Department of Geography, University of British Columbia, Vancouver, 2005.

Martín Mendiluce, J. "Los embalses en España: Su necesidad y trascendencia económica." *Revista de obras públicas* 3354 (1996): 7–24.

Mason, E., and R. Asher. *The World Bank since Bretton Woods*. Washington, DC: Brookings Institution, 1973.

McCarthy, J. "Commons as Counter-Hegemonic Projects." *Capitalism Nature Socialism* 16, no. 1 (2005): 9–24.

McCay, B.J., and S. Jentoft. "Market or Community Failure? Critical Perspectives on Common Property Research." *Human Organization* 57, no. 1 (1998): 21–29.

McCully, P. *Silenced Rivers: The Ecology and Politics of Large Dams*. London: Zed Books, 1996.

McDonald, D. "No Money, No Service: South Africa's Poorest Citizens Lose Out under Attempts to Recover Service Costs for Water and Power." *Alternatives* 28, no. 2 (2002a): 16–20.

McDonald, D., ed. *Environmental Justice in South Africa*. Athens and Cape Town: Ohio University Press and University of Cape Town Press, 2002b.

McDonald, D., and G. Ruiters. *The Age of Commodity: Water Privatization in Southern Africa*. London: Earthscan, 2005.

McGranahan, G., P. Jacobi, J. Songsore, C. Surjadi, and M. Kjellén. *The Citizens at Risk: From Urban Sanitation to Sustainable Cities*. London: Earthscan, 2001.

McIntosh, A. *Asian Water Supplies: Reaching the Urban Poor*. Manila: Asian Development Bank and International Water Association, 2003.

McKibben, B. *The End of Nature*. New York: Random House, 1989.

Megginson, W., and J. Netter. "From State to Market: A Survey of Empirical Studies on Privatization." *Journal of Economic Literature* 39, no. 2 (2001): 321–389.

Mehrotra, S., and E. Delamonica. "The Private Sector and Privatization in Social Services: Is the Washington Consensus Dead?" *Global Social Policy* 5, no. 2 (2005): 141–174.

Mehta, L. "Water, Difference, and Power: Unpacking Notions of Water 'Users' in Kutch, India." *International Journal of Water* 1, no. 3/4 (2001): 324–342.

———. *The Politics and Poetics of Water: Naturalising Scarcity in Western India*. London: Orient Longman, 2005.

Mehta, L., M. Leach, and I. Scoones. "Environmental Governance in an Uncertain World." *IDS Bulletin* 32, no. 4 (2001): 1–9.

Meier, G. *Biography of a Subject: An Evolution of Development Economics*. New York: Oxford University Press, 2005.

Meillassoux, C. *Maidens, Meal, and Money: Capitalism and the Domestic Community*. Cambridge: Cambridge University Press, 1981.

Melosi, M. "Pure and Plentiful: The Development of Modern Waterworks in the United States." *Water Policy* 2 (2000a): 243–265.

———. *The Sanitary City: Urban Infrastructure in America from Colonial Times to the Present*. Baltimore: Johns Hopkins University Press, 2000b.

Menegat, R. "Participatory Democracy and Sustainable Development: Integrated Urban Environmental Management in Porto Alegre, Brazil." *Environment and Urbanization* 14, no. 2 (2002): 181–206.

Mertes, T., ed. *A Movement of Movements: Is Another World Really Possible?* London: Verso, 2004.

Miller-Adams, M. *The World Bank: New Agendas in a Changing World*. New York: Routledge, 1999.

Millward, R. "Privatization in Historical Perspective: The UK Water Industry." In *Money, Trade and Payments: Essays in Honour of D.J. Coppock,* edited by D. Cobham, R. Harrington and G. Zis, 188–209. Manchester, UK: Manchester University Press, 1989.

Milone, P. "Indische Culture, and Its Relationship to Urban Life." *Comparative Studies in Society and History* 9 (1967): 407–426.

MIMAM. *Libro blanco del agua en España*. Madrid: Ministerio de Medio Ambiente, 1998.

Mitchell, B. "Integrated Water Resource Management, Institutional Arrangements, and Land-Use Planning." *Environment and Planning A* 37, no. 8 (2005): 1335–1352.

Mitchell, D. *The Right to the City: Social Justice and the Fight for Public Space*. London: Guilford Press, 2003.

Mitchell, T. *Rule of Experts: Egypt, Techno-Politics, Modernity*. Berkeley and Los Angeles: University of California Press, 2002.

Moens Bernelot, J.C. "Het drinkwater te Batavia: Geschiedenis van het drinkwater te Batavia, rapport omtrent een ingesteld onderzoek" [The Drinking Water in Batavia: History of the Drinking Water in Batavia, a Research Report]. *Geneeskundig tijdschif voor Nederlandsch-Indie* 15 (1873): 275–488.

Mol, A. "Ecological Modernisation and Institutional Reflexivity: Environmental Reform in the Late Modern Age." *Environmental Politics* 5, no. 2 (1996): 302–323.

Morgan, B. "The Regulatory Face of the Human Right to Water." *Journal of Water Law* 15, no. 5 (2004a): 179–186.

——. "Water: Frontier Markets and Cosmopolitan Activism." *Soundings: A Journal of Politics and Culture* 27 (2004b): 10–24.

——. "Social Protest against Privatization of Water: Forging Cosmopolitan Citizenship?" In *Sustainable Justice: Reconciling International Economic, Environmental and Social Law,* edited by M.-C. Cordonier Seggier and Judge C.G. Weeramantry, 339–352. Boston: Martinus Nijhoff, 2005.

——. "Consuming without Paying: Stealing or Campaigning? The Civic Implications of Civil Disobedience around Access to Water." In *Citizenship and Consumption,* edited by Kate Soper and Frank Trentmann. 71–86. Basingstoke: Palgrave Macmillan, 2008.

Morse, B. "Sardar Sarovar: Report of the Independent Review." Ottawa: Resources Futures International, 1992.

Mosse, D. "The Symbolic Making of a Common Property Resource: History, Ecology and Locality in a Tank Irrigated Landscape in South India." *Development and Change* 28, no. 3 (1997): 467–504.

——. *The Rule of Water: Statecraft, Ecology, and Collective Action in South India*. Oxford and New York: Oxford University Press, 2003.

Mrazek, R. *Engineers of Happyland: Technology and Nationalism in a Colony*. Princeton, NJ: Princeton University Press, 2002.

Mugabi, J., S. Kayaga, and I. Smout. "Why Water Utility Customers Don't Pay Their Bills Promptly." *Water Utility Management International* 2, no. 2 (2007): 17–19.

Mugero, C., and A. Hoque. "Review of Cholera Epidemic in South Africa, with Focus on KwaZulu-Natal Province. August 2000–11 April 2001." KwaZulu-Natal Department of Health, 2001.

Mujianto. "A Piece of Story from Ciliwung River Bank, Never Drink River Water." *PERCIK* 2004, 22–23.

Mukherjee. N. "Voices of the Poor: Making Services Work for the Poor in Indonesia; A Qualitative Consultation with the Poor at Eight Sites." Jakarta: World Bank, East Asia and Pacific Region, 2006.

Muller, M. "Parish Pump Politics: The Politics of Water Supply in South Africa." *Progress in Development Studies* 7, no. 1 (2007): 33–45.

——. "Free Basic Water: A Sustainable Instrument for a Sustainable Future in South Africa." *Environment and Urbanization* 20, no. 1 (2008): 67–87.

Mutua, M. *Human Rights: A Political and Cultural Critique.* Philadelphia: University of Pennsylvania Press, 2002.

Muzzini, E. *Consumer Participation in Infrastructure Regulation: Evidence from the East Asia and Pacific Region.* Washington, DC: World Bank, 2005.

Narain, S. "Community-Led Alternatives to Water Management: India Case Study." Background paper, Human Development Report. New York: United Nations Development Programme, 2006.

National Research Council. "Privatization of Water Services in the United States: An Assessment of Issues and Experience." Washington, DC: National Academy Press, 2002.

Nellis, J. "Time to Rethink Privatization in Transition Economies?" *Finance & Development* 36, no. 2 (1999): 16–19.

——. "The World Bank, Privatization, and Enterprise Reform in Transition Economies: A Retrospective Analysis." Washington, DC: World Bank, Operations and Evaluation Department, 2002.

Newbery, D. *Privatization, Restructuring, and Regulation of Network Utilities.* Cambridge, MA: MIT Press, 2000.

Newson, M. *Land, Water and Development.* 2nd ed. London: Routledge, 1997.

Nickson, A. "The Public-Private Mix in Urban Water Supply." *International Review of Administrative Sciences* 63, no. 2 (1997): 165–186.

Nickson, A., and R. Franceys. *Tapping the Market: The Challenge of Institutional Reform in the Urban Water Sector.* Basingstoke: Palgrave Macmillan, 2003.

Nickson, A., and C. Vargas. "The Limitations of Water Regulation: The Failure of the Cochabamba Concession in Bolivia." *Bulletin of Latin American Research* 21, no. 1 (2002): 99–120.

Norman, E., and K. Bakker. "Transgressing Scales: Transboundary Water Governance across the Canada-US Borderland." *Annals of the Association of American Geographers* 99, no. 1 (2009), 99–117.

North, D. "The New Institutional Economics and Third World Development." In *The New Institutional Economics and Third World Development,* edited by J. Harriss, J. Hunter, and C. Lewis, 17–26. London: Routledge, 1995.

Nurbianto, B. "New Investors, Old Problems in Water Industry." *Jakarta Post,* January 19, 2007.

O'Connor, J. *Fiscal Crisis of the State.* New York: St. Martin's, 1973.

——. "The Second Contradiction of Capitalism." In *The Greening of Marxism,* edited by T. Benton, 197–221. London: Guilford, 1996.

OECD. "DAC Orientations for Development Cooperation in Support of Private Sector Development." Paris: Organisation for Economic Cooperation and Development, 1995.

——. *Global Trends in Urban Water Supply and Waste Water Financing and Management: Changing Roles for the Public and Private Sectors.* Paris: Organisation for Economic Cooperation and Development, 2000.

——. *Managing Water for All: An OECD Perspective on Pricing and Financing.* Paris: OECD, 2009.

OECD-DAC. "Promoting Private Investment for Development: The Role of ODA." Paris: Organisation for Economic Cooperation and Development, Development Assistance Committee, 2006.

Ofwat. "The Proposed Restructuring of the Kelda Group: A Preliminary Assessment by the Director General of Water Services, Office of Water Services." Birmingham, UK: Ofwat, 2000.

Ohlsson, L. "Water Conflicts and Social Resource Scarcity." *Physics and Chemistry of the Earth (B)* 25, no. 3 (2000): 213–220.

Olivera, O. "The Rising of the Waters: The 'Commons' Re-established by the People of Bolivia." Unpublished manuscript. Cochabamba, Bolivia: Octubre Azul, n.d.

Olivera, O., and T. Lewis. *Cochabamba! Water War in Bolivia.* Cambridge, MA: South End Press, 2004.

O'Riordan, T. "Environmental Science, Sustainability and Politics." *Transactions of the Institute of British Geographers* 29, no. 2 (2004): 234–247.

Ortolano, L., and K. Cushing. "Grand Coulee Dam 70 Years Later: What Can We Learn?" *International Journal of Water Resources Development* 18, no. 3 (September 2002), 373–390.

Osterweil, M. "Place-Based Globalism: Theorizing the Global Justice Movement." *Development* 48, no. 2 (2005): 23–28.

Ostrom, E. *Governing the Commons: The Evolution of Institutions for Collective Action.* New York: Cambridge University Press, 1990.

Ostrom, E., and R. Keohane, eds. *Local Commons and Global Interdependence: Heterogeneity and Cooperation in Two Domains.* London: Sage, 1995.

Ostrom, E., L. Schroeder, and S. Wynne. *Institutional Incentives and Sustainable Development: Infrastructure Policies in Perspective.* Boulder, CO: Westview Press, 1993.

Page, B. "Paying for Water and the Geography of Commodities." *Transactions of the Institute of British Geographers* 30, no. 3 (2005), 293–306.

Page, B., and K. Bakker. "Water Governance and Water Users in a Privatized Water Industry: Participation in Policy-Making and in Water Services Provision—a Case Study of England and Wales." *International Journal of Water* 3, no. 1 (2005): 38–60.

Palaniappan, M., H. Cooley, P. Gleick, and G. Wolff. "Assessing the Long-Term Outlook for Current Business Models in the Construction and Provision of Water Infrastructure and Services." Pacific Institute, Report authored for the Organisation for Economic Cooperation and Development, 2006.

PALYJA. *Visi, Misi, Nilai* [Vision, Mission, Values]. Public brochure. Jakarta: Palyja/Suez, 2008.

PAM Jaya. "PAM Jaya 70 Years, 1922–1992." Jakarta, Indonesia: PAM Jaya, 1992a.

——. "Usaha peningkatan pelayanan masyarakat" [Efforts to Improve Services for the Public]. Jakarta, Indonesia: PAM Jaya, 1992b.

——. *Instalasi produksi pulogadung* [Production Installation Pulogadung]. Jakarta: PAM Jaya, 1992c.

——. "Laporan evaluasi kinerja tahunan." Jakarta: Perusahaan Daerah Air Mimun, 2004.

Pandjaitan, G. "Kondisi pelayanan air minum di Dki Jakarta sebelum swastanisasi" [Condition Service of Drinking Water in Dki Jakarta before Privatization]. Paper presented at the Workshop on Drinking Water in Jakarta. Jakarta, Indonesia, August 2004.

Paterson, C., D. Mara, and T. Curis. "Pro-Poor Sanitation Technologies." *Geoforum* 38, no. 5 (2007): 901–907.

Payen, G. "Statement on the Right to Water and Role of Local Governments." Aquafed, 2006. http://www.aquafed.org/pdf/WWF4-opening_GP_RTW-LocGov_Pc_2006–03-16.pdf.

Peprah, I. "The African Development Bank: Taking Stock and Preparing for the 21st Century." Ottawa: C-C Consulting, 1994.

Perelman, M. *The Invention of Capitalism: Classical Political Economy and the Secret History of Primitive Accumulation.* Durham, NC: Duke University Press, 2000.

PERPAMSI. "Air minum" [Drinking Water]. *WARTA Journal PERPAMSI* 1, no. 1–2 (1975): 34.

Perreault, T. "State Restructuring and the Scale Politics of Rural Water Governance in Bolivia." *Environment and Planning A* 37, no. 2 (2005): 263–284.

——. "From the Guerra Del Agua to the Guerra Del Gas: Resource Governance, Popular Protest and Social Justice in Bolivia." *Antipode* 38, no. 1 (2006): 150–172.

——. "Custom and Contradiction: Rural Water Governance and the Politics of *Usos y Costumbres* in Bolivia's Irrigators' Movement." *Annals of the Association of American Geographers* 98, no. 4 (2008), 834–854.

Petrella, R. *The Water Manifesto: Arguments for a World Water Contract.* London: Zed Books, 2001.

Pierre, J. "The Marketization of the State: Citizens, Customers and the Emergence of the Public Market." In *Governance in a Changing Environment,* edited by G. Peters and D. Savoie, 55–81. Montreal and Kingston: McGill-Queens University Press, 1995.

——, ed. *Debating Governance.* Oxford: Oxford University Press, 2000.

Pierre, J., and B. Peters. *Governance, Politics and the State.* Basingstoke: Palgrave Macmillan, 2000.

Pitman, G.K. "Bridging Troubled Waters: Assessing the World Bank Water Resources Strategy." Washington, DC: World Bank, 2002.

Platteau, J., and F. Gaspart. "The Risk of Resource Misappropriation in Community-Driven Development." *World Development* 31, no. 10 (2003): 1687–1703.

Platts Global Water Report. "Jakarta: Test Bed for Public Private Partnership." Issue 146, 1–3, 2002.

Polanyi, K. *The Great Transformation: The Political and Economic Origins of Our Time.* New York: Farrar & Reinhart, 1944.

Ponniah, T. "Democracy vs. Empire: Alternatives to Globalization Presented at the World Social Forum." *Antipode* 36, no. 1 (2004): 130–133.

Ponniah, T., and W. Fisher, eds. *Another World Is Possible: Popular Alternatives to Globalization at the World Social Forum.* New York: Zed Press, 2003.

Porter, R. *The Economics of Water and Waste: A Case Study of Jakarta, Indonesia.* Aldershot: Avebury, 1996.

Postel, S. *Pillar of Sand: Can the Irrigation Miracle Last?* New York: Norton, 1999.

Potkanski, T., and W. Adams. "Water Scarcity, Property Regimes and Irrigation Management in Sonjo, Tanzania." *Journal of Development Studies* 34, no. 4 (1998): 86–116.

Prasad, N. "Privatisation Results: Private Sector Participation in Water Services after 15 Years." *Development Policy Review* 24, no. 6 (2006): 669–692.

PricewaterhouseCoopers. "Comparative Review of IFI Risk Mitigation Instruments and Direct Sub-Sovereign Lending." London: PricewaterhouseCoopers Securities LLC, World Bank, and Bank-Netherlands Water Partnership, 2003.

——. "Hybrid PPPs: Levering EU Funds and Private Capital." Washington, DC: Public Private Infrastructure Advisory Facility and World Bank, 2006.

Prudham, S. "Poisoning the Well: Neoliberalism and the Contamination of Municipal Water in Walkerton, Ontario." *Geoforum* 35, no. 3 (2004): 343–359.

PSIRU. "Public-Public Partnerships as a Catalyst for Capacity Building and Institutional Development: Lessons from Stockholm Vatten's Experience in the Baltic Region." University of Greenwich, Public Services International Research Unit, 2006.

Public Citizen. "Public-Public Partnerships: A Backgrounder on Successful Water/Wastewater Reengineering Programs." Washington, DC: Public Citizen and Food and Water Watch, 2002.

Putnam, R. *Bowling Alone: The Collapse and Revival of American Community.* New York: Simon & Schuster, 2000.

Raco, M. "Assessing Community Participation in Local Economic Development: Lessons for the New Urban Policy." *Political Geography* 19, no. 5 (2000): 573–599.

Raco, M., and J. Flint. "Communities, Places and Institutional Relations: Assessing the Role of Area-Based Community Representation in Local Governance." *Political Geography* 20, no. 5 (2001): 585–612.

Rao, V. "Slum as Theory: The South/Asian City and Globalization." *International Journal of Urban and Regional Research* 30, no. 1 (2006): 225–232.

Ratner, B.D. "Reasserting Community: The Social Challenge of Wastewater Management in Panajachel, Guatemala." *Human Organization* 63, no. 1 (2004): 47–56.

Razzaque, J. "Trading Water: The Human Factor." *Review of European Community and International Environmental Law* 13, no. 1 (2004): 15–26.

Renzetti, S., and D. Dupont. "The Performance of Municipal Water Supply Utilities: Evidence on the Role of Ownership." *Journal of Toxicology and Environmental Health, Part A* 67, no. 20–22 (2004): 1861–1878.

Rhodes, R. "The New Governance: Governing without Government." *Political Studies* 44, no. 3 (1996): 652–667.

Rice, X. "The Water Margin." *Guardian*, August 16, 2007.

Rich, B. *Mortgaging the Earth: The World Bank, Environmental Impoverishment and the Crisis of Development*. London: Earthscan, 1994.

Ringskog, K., M. Hammond, and A. Locussol. "Using Management and Lease-Affermage Contracts for Water Supply: How Effective Are They?" *Gridlines* 12. Washington, DC: World Bank, 2006.

Rivera, D. "Private Sector Participation in Water Supply and Sanitation: Lessons from Six Developing Countries." Washington, DC: World Bank, 1996.

Rivera, J. *Acequia Culture: Water, Land and Community in the Southwest*. Albuquerque: University of New Mexico Press, 1998.

Robbins, P. "Transnational Corporations and the Discourse of Water Privatization." *Journal of International Development* 15, no. 8 (2003): 1073–1082.

Roberts, R., and J. Emel. "Groundwater Management in the Southern High Plains: Questioning the Tragedy of the Commons." *Economic Geography* 68, no. 3 (1992): 249–271.

Robertson, M. "No Net Loss: Wetland Restoration and the Incomplete Capitalization of Nature." *Antipode* 32, no. 4 (2000): 463–493.

——. "The Neoliberalization of Ecosystem Services: Wetland Mitigation Banking and Problems in Environmental Governance." *Geoforum* 35, no. 3 (2004): 361–373.

——. "Discovering Price in All the Wrong Places: The Work of Commodity Definition and Price under Neoliberal Environmental Policy." *Antipode* 39, no. 3 (2007): 500–526.

Robertson-Snape, F. "Corruption, Collusion and Nepotism in Indonesia." *Third World Quarterly* 20, no. 3 (1999): 589–602.

Robinson, J. *Ordinary Cities: Between Modernity and Development*. London: Routledge, 2006.

Roger, N. "Recent Trends in Private Participation in Infrastructure." *Private Sector Viewpoint* 196. Washington, DC: World Bank, 1999.

Rogers, P. *America's Water: Federal Roles and Responsibilities*. Cambridge, MA: MIT Press, 1993.

——. "America's Water: Federal Roles and Responsibilities." *Water Policy* 4, no. 5 (2003): 423–426.

Rogers, P., and A. Hall. "Effective Water Governance." TEC Background Papers, no. 7. Sweden: Global Water Partnership Technical Committee, 2003.

Rogers, P., R. Silva, and R. Bhatia. "Water Is an Economic Good: How to Use Prices to Promote Equity, Efficiency, and Sustainability." *Water Policy* 4, no. 1 (2002): 1–17.

Rondinelli, D.A., and G.S. Cheema. *Urban Services in Developing Countries: Public and Private Roles in Urban Development*. London: Macmillan, 1988.

Rorty, R. "Human Rights, Rationality, and Sentimentality." In *On Human Rights: The Oxford Amnesty Lectures*, edited by S. Shute and S. Hurley, 112–134. New York: Basic Books, 1993.

Rostow, W. *The Stages of Economic Growth*. Cambridge: Cambridge University Press, 1960.

Roth, D., R. Boelens, and M. Zwarteveen. *Liquid Relations: Contested Water Rights and Legal Complexity*. New Brunswick, NJ: Rutgers University Press, 2005.

Rowe, J. "Parallel Economy of the Commons." In *State of the World Report 2008: Innovations for a Sustainable Economy*, 138–150. Washington, DC: Worldwatch Institute, 2008.

Roy, A. "Urban Informality: Toward an Epistemology of Planning." *Journal of the American Planning Association* 71, no. 2 (2005): 147.

Ruiz-Meir, F., and M. van Ginneken. *Consumer Cooperatives: An Alternative Institutional Model for Delivery of Urban Water Supply and Sanitation Services?* Water Supply and Sanitation working note no. 5. Washington, DC: Water Supply and Sanitation Sector Board, Infrastructure Network, World Bank Group, 2006.

Saad-Filho, A., and D. Johnston. *Neoliberalism: A Critical Reader*. London: Pluto Press, 2005.

Sabarini, P. "Clean Water Luxury for North Jakarta Slum Dwellers." *Jakarta Post*, January 27, 2007.

Sabatier, P., W. Focht, M. Lubell, Z. Trachtenberg, A. Vedlitz, and M. Matlock, eds. *Swimming Upstream: Collaborative Approaches to Watershed Management*. Cambridge, MA: MIT Press, 2005.

Sachs, J. "Lessons of Transition: A Reform Architect Looks at What Worked—and What Didn't." *Central European Economic Review* October 25, 1999.

Sadove, R., and P.J. Squire. "The Bank's Policy towards Municipal Supply Projects." Washington, DC: World Bank, 1960.

Saleth, R., and A. Dinar. "Evaluating Water Institutions and Water Sector Performance." World Bank Technical Paper no. 447. Washington, DC: World Bank, 1999.

Salman, S., and S. McInerney-Lankford. *The Human Right to Water: Legal and Policy Dimensions*. Washington, DC: World Bank, 2004.

Sappington, D., and J. Stiglitz. "Information and Regulation." In *Public Regulation: New Perspectives on Institutions and Policies*, edited by E. Bailey, 3–43. Cambridge, MA: MIT Press, 1987a.

——. "Privatization, Information, and Incentives." *Journal of Policy Analysis and Management* 6, no. 3 (1987b): 567–582.

Sawkins, J. "Yardstick Competition in the English and Welsh Water Industry: Fiction or Reality?" *Utilities Policy* 5, no. 1 (1995): 27–36.

Schreiber, W. "Realizing the Right to Water in International Investment Law: An Interdisciplinary Approach to BIT Obligations." *Natural Resources Journal* 48, no. 2 (2008): 431–478.

Schultz, J. "Water in Cochabamba after the Water Revolt: A Legend with Mixed Results." In *Dignity and Defiance: Stories from Bolivia's Challenge to Globalization*, edited by J. Shultz and M. Draper, 9–34. Berkeley: University of California Press, 2008.

Scott, J. "The Analysis of Corruption in Developing Nations." *Comparative Studies in Society and History* 11, no. 3 (1969): 315–341.

——. *Seeing Like a State: How Certain Schemes to Improve the Human Condition Have Failed*. New Haven, CT: Yale University Press, 1998.

Scottish Parliament. "Water Industry (Scotland) Bill: Policy Memorandum." Edinburgh: The Stationery Office, 2001.

Scudder, T. *The Future of Large Dams: Dealing with Social, Environmental, Institutional and Political Costs*. London: Earthscan, 2005.

Segerfeldt, F. *Water for Sale: How Business and the Market Can Resolve the World's Water Crisis*. Washington, DC: Cato Institute, 2005.

Sen, A. *Inequality Reexamined*. Cambridge, MA: Harvard University Press, 1992.

——. *Commodities and Capabilities*. Oxford: Oxford University Press, 1999.

Sheail, J. "Planning, Water Supplies and Ministerial Power in Inter-War Britain." *Public Administration* 61, no. 4 (1983): 386–395.

——. "Government and the Perception of Reservoir Development in Britain: An Historical Perspective." *Planning Perspectives* 1, no. 1 (1986): 45–60.

Shiklomanov, I.A. "Comprehensive Assessment of the Freshwater Resources of the World: Assessment of Water Resources and Water Availability in the World." Stockholm: World Meteorological Organization and Stockholm Environment Institute, 1997.

Shirley, M. *Thirsting for Efficiency: The Economics and Politics of Urban Water System Reform*. Washington, DC, World Bank, 2002.

Shirley, M., and P. Walsh "Public versus Private Ownership: The Current State of the Debate." Washington, DC, World Bank, Development Research Group, Regulation and Competition Policy, 2000.

Shiva, V. *Water Wars: Privatization, Pollution and Profit*. London: Pluto Press, 2002.

Shofiani, N. "Reconstruction of Indonesia's Drinking Water Utilities: Assessment and Stakeholders Perspective of Private Sector Participation in the Capital Province of Jakarta, Indonesia." M.Sc. thesis, Royal Institute of Technology, 2003.

Silas, J. "Government-Community Partnerships in Kampung Improvement Programmes in Surabaya." *Environment and Urbanization* 4, no. 2 (1992): 33–41.

Silva, G., N. Tynan, and Y. Yilmaz. "Private Participation in the Water and Sewerage Sector: Recent Trends." In *Private Sector Viewpoint*, note no. 147, 1–8. World Bank—Finance, Private Sector, and Infrastructure Network, 1998.

Simamora, A. "Groundwater Pollution 'Worse Than Thought.'" *Jakarta Post*, January 19, 2007.

Simanjuntak, C.H., W. Larasati, S. Arjoso, M. Putri, M. Lesmana, B.A. Oyofo, N. Sukri, et al. "Cholera in Indonesia in 1993–1999." *American Journal of Tropical Medicine and Hygiene* 65, no. 6 (2001): 788–797.

Smets, H. "Diluted View of Water as a Right: 4th World Water Forum." *Environmental Policy and Law* 36, no. 2 (2006a): 88–93.

——. "The Right to Water in National Legislation." Paris: Agence Française de Développement, 2006b.

Smith, N. *Uneven Development: Nature, Capital and the Production of Space*. New York: Blackwell, 1984.

Smitt, A. *De waterleiding van Batavia* [The Water Network of Batavia]. Batavia: Landsdrukkerij, 1922.

Sokile, C.S., and B. van Koppen. "Local Water Rights and Local Water User Entities: The Unsung Heroines to Water Resource Management in Tanzania." *Physics and Chemistry of the Earth* 29 (2004): 1349–1356.

Solley, W., R. Pierce, and H. Perlman. "Estimated Use of Water in the United States in 1995." In *United States Geological Survey Circular 1200*. Reston, VA: U.S. Department of the Interior, 1998.

Solo, T. "Small-Scale Entrepreneurs in the Urban Water and Sanitation Market." *Environment and Urbanization* 11, no. 1 (1999): 117–132.

Sparke, M. "Political Geography: Political Geographies of Globalization (2); Governance." *Progress in Human Geography* 30, no. 3 (2006): 357–372.

Spiller, P., and W. Savedoff. "Government Opportunism and the Provision of Water." In *Spilled Water: Institutional Commitment in the Provision of Water Services,* edited by P. Spiller and W. Savedoff. Washington, DC: Inter-American Development Bank, 1999.

Spronk, S. "Roots of Resistance to Urban Water Privatization in Bolivia: Public Services, the 'New Working Class' and the Crisis of Neoliberalism." *International Labor and Working Class History* 71 (2007): 8–28.

Spronk, S., and C. Flores. *Después de las guerras del agua.* La Paz: Plural Editores, 2007.

——. "Water, National Sovereignty and Social Resistance: Bilateral Investment Treaties and the Struggles against Multinational Water Companies in Cochabamba and El Alto, Bolivia." *Law, Social Justice and Global Development* 1 (2008).

Squatriti, P. *Water and Society in Early Medieval Italy, A.D. 400–1000.* Cambridge: Cambridge University Press, 1998.

Staples, A. *The Birth of Development: How the World Bank, Food and Agriculture Organization, and World Health Organization Changed the World, 1945–1965.* Kent, OH: Kent State University Press, 2006.

Stewart, A., and T. Gray. "The Authenticity of 'Type Two' Multistakeholder Partnerships for Water and Sanitation in Africa: When Is a Stakeholder a Partner?" *Environmental Politics* 15, no. 3 (2006): 362–378.

Stiglitz, J. *Whither Socialism?* Cambridge, MA: MIT Press, 1994.

——. "More Instruments and Broader Goals: Moving toward the Post-Washington Consensus." Helsinki: World Institute for Development Economics Research, 1998.

——. *Globalization and Its Discontents.* New York: W.W. Norton, 2003.

Stigson, B. "What Is the Role of the Private Sector in Sustainable Management of · Water?" In *Water and International Solidarity: The Geneva Questions—Round Table,* edited by Mahmoud Abu-Zeid, Kaj Bärlund, Wilfried Kreisel, Philippe Roch, Ismail Serageldin, and Björn Stigson, 34–39. Geneva: Green Cross International, 1997.

St. Martin, K. "Disrupting Enclosure in the New England Fisheries." *Capitalism Nature Socialism* 16, no. 1 (2005): 63–80.

Stoler, A. "Sexual Affronts and Racial Frontiers: European Identities and the Cultural Politics of Exclusion in Colonial Southeast Asia." *Comparative Studies in Society and History* 34, no. 3 (1992): 514–551.

——. *Race and the Education of Desire: Foucault's History of Sexuality and the Colonial Order of Things.* Durham, NC: Duke University Press, 1995.

Strang, V. *The Meaning of Water.* Oxford: Berg, 2004.

Surjadi, C. "Public Private Partnerships and the Poor: Drinking Water Concessions." In *A Study for Better Understanding Public-Private Partnerships and Water Provision in Low-Income Settlements.* Leicestershire: Water, Engineering and Development Centre, Loughborough University, 2003.

Surjadi, C., L. Padmasutra, D. Wahyuningsih, G. McGranahan, and M. Kjellén. *Household Environmental Problems in Jakarta.* Stockholm: Stockholm Environment Institute, 1994.

Susantono, B. "Informal Water Services in Metropolitan Cities of the Developing World: The Case of Jakarta, Indonesia." PhD diss., University of California–Berkeley, 2001.

Susilowati, I., and L. Budiata. "An Introduction of Co-Management Approach into Babon River Management in Semarang, Central Java, Indonesia." *Water Science and Technology* 48, no. 7 (2003): 173–180.

Swyngedouw, E. "Power, Nature, and the City: The Conquest of Water and the Political Ecology of Urbanization in Guayaquil, Ecuador; 1880–1990." *Environment and Planning A* 29, no. 2 (1997): 311–332.

———. "Modernity and Hybridity: Nature, *Regeneracionismo*, and the Production of the Spanish Waterscape, 1890–1930." *Annals of the Association of American Geographers* 89, no. 3 (1999): 443–465.

———. *Social Power and the Urbanization of Water: Flows of Power.* Oxford: Oxford University Press, 2004.

———. "Dispossessing H2O: The Contested Terrain of Water Privatization." *Capitalism, Nature, Socialism* 16, no. 1 (2005): 81–98.

Takahashi, K. "Globalization and Management of Water Resources: Development Opportunities and Constraints of Diversified Developing Countries." *International Journal of Water Resources Development* 17, no. 4 (2001): 481–487.

Tarr, J. *The Search for the Ultimate Sink: Urban Pollution in Historical Perspective.* Akron, OH: University of Akron Press, 1996.

Tarr, J., and G. Dupuy, eds. *Technology and the Rise of the Networked City in Europe and America.* Philadelphia: Temple University Press, 1988.

Taylor, C. *Modern Social Imaginaries.* Durham, NC and London: Duke University Press, 2004.

Taylor, J. "An Evaluation of Selected Impacts of Jakarta's Kampung Improvement Program." PhD diss., University of California–Los Angeles, 1983.

Taylor, R. *Creating a Better World: Interpreting Global Civil Society.* Bloomfield, CT: Kumarian, 2004.

Tendler, J. *Good Government in the Tropics.* Baltimore: Johns Hopkins University Press, 1997.

Thompson, E.P. "The Moral Economy of the English Crowd in the Eighteenth Century." *Past & Present* 50 (1971): 76–136.

Tissington, K., M. Dettmann, M. Langford, J. Dugard, and S. Conteh. *Water Services Fault Lines: An Assessment of South Africa's Water and Sanitation Provision across 15 Municipalities.* Johannesburg: Centre for Applied Legal Studies, 2008.

Travis, R., J. Morris, and E. Morris. "State Implementation of Federal Environmental Policy: Explaining Leveraging in the Clean Water State Revolving Fund." *Policy Studies Journal* 32, no. 3 (2004): 461–480.

Trawick, P. "Moral Economy of Water: Equity and Antiquity in the Andean Commons." *American Anthropologist* 103, no. 2 (2001a): 361–379.

———. "Successfully Governing the Commons: Principles of Social Organization in an Andean Irrigation System." *Human Ecology* 29, no. 1 (2001b): 1–25.

———. *The Struggle for Water in Peru: Comedy and Tragedy in the Andean Commons.* Stanford: Stanford University Press, 2003.

Tussie, D. *The Inter-American Development Bank.* Boulder, CO: Lynne Rienner, 1995.

Tutuko, K. "Jakarta Water Supply." Sustainable Urban Services, Hong Kong Seminar. Hong Kong: Pam Jaya, 2001.

Ugaz, C. "Consumer Participation and Pro-Poor Regulation in Latin America." Wider Discussion Paper 2002/121. Helsinki: World Institute for Development Economics Research, 2002.

Uijterlinde, R., A. Janssen, and C. Figueres, eds. "Success Factors in Self-Financing Local Water Management: A Contribution to the Third World Water Forum in Japan, 2003." Amsterdam: Partners for Water, 2003.

UNCHS. *Water and Sanitation in the World's Cities: Local Action for Global Goals.* London: Earthscan, 2003.

——. *Meeting Development Goals in Small Urban Centres: Water and Sanitation in the World's Cities 2006.* London: Earthscan, 2006.

UNCITRAL. *City Water Services Ltd. v. Dar es Salaam Water and Sewerage Authority (UNCITRAL Arbitration, case no. UN6761).* Vienna: United Nations Commission on International Trade Law, 2007.

UNCSD. "Report on the Thirteenth Session." E/CN.17/2005/12. New York: UN Commission on Sustainable Development, 2005.

UNDP. *Human Development Report 2000: Human Rights and Human Development.* New York: United Nations Development Programme, Human Development Report Office, 2000.

——. *Human Development Report 2003: Millennium Development Goals; A Compact for Nations to End Human Poverty.* New York: United Nations Development Programme, Human Development Report Office, 2003.

——. *Human Development Report 2005: International Cooperation at a Crossroads: Aid, Trade and Security in an Unequal World.* New York: United Nations Development Programme, Human Development Report Office, 2005.

——. *Human Development Report 2006: Beyond Scarcity; Power, Poverty, and the Global Water Crisis.* New York: United Nations Development Programme, Human Development Report Office, 2006.

——. *The Global UNDP Water Governance Strategy: A Follow-Up to Human Development Report 2006; Beyond Scarcity: Power, Poverty and the Global Water Crisis.* New York: United Nations Development Programme, 2007.

UN ECOSOC. "Economic, Social and Cultural Rights: The Right to Food." Report submitted to the 59th session of the Commission on Human Rights, by the Special Rapporteur on the Right to Food, Jean Ziegler, in accordance with Commission on Human Rights resolution 2002/25. E/CN.4/2003/54. UN Economic and Social Council, 2003.

UNFPA. *State of the World Population Report 2008: Reaching Common Ground; Culture, Gender, and Human Rights.* Geneva: UN Population Fund, 2008.

United States General Accounting Office. "Water Infrastructure: Information on Financing, Capital Planning, and Privatization." Report to Congressional Requesters GAO-02-764. Washington, DC: United States General Accounting Office, 2002.

UN Millennium Project. *Investing in Development: A Practical Plan to Achieve the Millennium Development Goals.* London: Earthscan, 2005.

UN-OHCHR. "Status of Ratifications of the Principal International Human Rights Treaties." Geneva: Office of the United Nations High Commissioner for Human Rights, 2002.

UNWWAP. "Water for People, Water for Life." Rome: FAO, 2003.

——. "Water: A Shared Responsibility." Rome: FAO, 2006.

Van Breen, H. "Beschouwingen: Van de technische-en watervoorzieningscommissies betreffende de verschillende in zake de voorgenomen verbetering van de watervoorziening der gemeente Batavia verschenen artikelen en uitgebrachte adviezen" [Reviews: Of the Technical and Water Supply Commissions Concerning the Different Articles and Advice Dealing with the Intended Improvement of Water Provision for the Municipality of Batavia]. Gemeenteraad van Batavia, gedrukte stukken no.40 [Municipal Council of Batavia, printed pieces no.40). Batavia: Ruygrok, 1916.

——. "Klein werken ter verberting van de gezondheidstoestand der hoofdplaats Batavia" [Small Works for Improvement of the Health Conditions in the Capital Batavia]. De waterstaats-ingenieur 3 (1919): 130–138.

Van De Giesen, N., M. Andreini, A. Van Edig, and P. Vlek. "Competition for Water Resources of the Volta Basin." Paper presented at the Regional Management of Water Resource, Maastrich, Netherlands, 2001.

Vandenhole, W., and T. Wielders. "Water as a Human Right: Water as an Essential Service: Does It Matter?" Netherlands Quarterly of Human Rights 26, no. 3 (2008): 391–424.

Van der Kop, G. "Under the Smoke of Java's Capital." Inter-Ocean 7, no. 3 (1926): 147–154.

Van der Kroef, J. Indonesia in the Modern World. Vol. 1. Bandung: Masa Baru, 1954.

Van der Merwe, S. "An Overview of Water Supply Management in South Africa." Aqua 44, no. 4 (1995): 151–160.

Van Doorne, J. A Divided Society: Segmentation and Mediation in Late-Colonial Indonesia. Rotterdam: Erasmus University, 1983.

Van Harten, G. "Private Authority and Transnational Governance: The Contours of the International System of Investor Protection." Review of International Political Economy 12, no. 4 (2005): 600–623.

Van Hofwegen, P. "Task Force on Financing Water for All: Enhancing Access to Finance for Local Governments, Financing Water for Agriculture. Report 1 (Gurria Report)." Marseilles: World Water Council, 2006.

Van Leeuwen, C. "Het huidige vraagstuk der watervoorzieining van Batavia en Mr. Cornelis: Voordracht, gehouden in de openbare vergadering der afdeeling Batavia van 14 Februari 1917." De waterstaats ingenieur 3 (1917): 81–103.

——. "Het rioleeringsvraagstuk in Nederlandsch-Indie" [The Question of Water Treatment in the Netherlands Indies]. De waterstaats ingenieur 5 (1920): 196–212.

Van Raay. "Watervoorziening van Batavia I" [Water Provision for Batavia, Part 1]. De ingenieur 7 (1915): 115–130.

Van Rooy, A. "The Frontiers of Influence: NGO Lobbying at the 1974 World Food Conference, the 1992 Earth Summit and Beyond." World Development 25, no. 1 (1997): 93–114.

———. *The Global Legitimacy Game: Civil Society, Globalization, and Protest.* New York: Palgrave Macmillan, 2004.

Verhoest, K., B. Peters, G. Bouckaert, and B. Verschuere. "The Study of Organizational Autonomy: A Conceptual Review." *Public Administration and Development* 24, no. 2 (2004): 101–118.

Vervoort, M. "Glimpses along Batavia's Canals." *Inter-Ocean* 7, no. 5 (1926): 265–272.

Vidal, J. "People Wake Up Angry at Being Alive in a Society Like This." *Guardian,* March 5, 2005.

Von Tunzelmann, N. *Steam Power and British Industrialization to 1860.* Oxford: Clarendon Press, 1978.

Wade, R. *Governing the Market: Economic Theory and the Role of Government in East Asian Industrialization.* Princeton, NJ: Princeton University Press, 1990.

———. *Village Republics: Economic Conditions for Collective Action in South India.* San Francisco: ICS Press, 1988.

Wallsten, S., and K. Kosec. *Public or Private Drinking Water? The Effects of Ownership and Benchmark Competition on U.S. Water System Regulatory Compliance and Household Water Expenditures.* AEI-Brookings Joint Center Working Paper no. 05–05, March 2005. http://ssrn.com/abstract=707131 or DOI: 10.2139/ssrn.707131.

Ward, C. *Reflected in Water: A Crisis in Social Responsibility.* London: Cassell, 1997.

Ward, F.A., and A. Michelsen. "The Economic Value of Water in Agriculture: Concepts and Policy Applications." *Water Policy* 4, no. 5 (2002): 423–446.

WaterAid, and Tearfund. "An Empty Glass: The EU Water Initiative's Contribution to the Water and Sanitation Millennium Targets." London: WaterAid and Tearfund, 2005.

WCD. *Dams and Development: A New Framework for Decision-Making; Report of the World Commission on Dams.* London: Earthscan, 2000.

WDM. "Dirty Aid, Dirty Water: The UK Government's Push to Privatise Water and Sanitation in Poor Countries." London: World Development Movement, 2005.

———. "Out of Time: The Case for Replacing the World Bank and IMF." London: World Development Movement, 2006.

———. "Going Public: Southern Solutions to the Global Water Crisis." London: World Development Movement, 2007.

Weber, S. "Origins of the European Bank for Reconstruction and Development." *International Organization* 48, no. 1 (1994): 1–38.

Weimer, M. "Action Research on Point of Use Drinking Water Treatment Alternatives as Appropriate for Underprivileged Households in Jakarta." Jakarta: USAID and Environmental Services Program, 2006.

Weintraub, J., and K. Kumar, eds. *Public and Private in Thought and Practice: Perspectives on a Grand Dichotomy.* Chicago: University of Chicago Press, 1997.

Weitz, A., and R. Franceys. *Beyond Boundaries, Extending Services to the Urban Poor.* Manila, Philippines: Asian Development Bank, 2002.

White, G.F., D.J. Bradley, and A.U. White. *Drawers of Water: Domestic Water Use in East Africa.* Chicago: University of Chicago Press, 1972.

Whittington, D. "Possible Adverse Effects of Increasing Block Water Tariffs in Developing Countries." *Economic Development and Cultural Change* 41, no. 1 (1992): 75–87.

WHO. "The International Drinking Water Supply and Sanitation Decade: End of Decade Review." Geneva: World Health Organization, 1992.

———. *Domestic Water Quantity, Service Level, and Health.* Geneva: World Health Organization, 2003a.

———. *The Right to Water.* Geneva: World Health Organization, 2003b.

———. "UN-Water Global Annual Assessment of Sanitation and Drinking Water. Pilot Report: Testing a New Reporting Approach." Geneva: World Health Organization, 2008.

WHO/UNICEF. "Meeting the MDG Water and Sanitation Target." Geneva: WHO/UNICEF Joint Monitoring Programme for Water Supply and Sanitation, 2006.

WHO, UNICEF, Water Supply and Sanitation Collaborative Council. *Global Water Supply and Sanitation Assessment 2000 Report.* Geneva: World Health Organization, 2000.

Wilder, M., and P.R. Lankao. "Paradoxes of Decentralization: Water Reform and Social Implications in Mexico." *World Development* 34, no. 11 (2006): 1977–1995.

Williamson, J. "What Should the World Bank Think about the Washington Consensus?" *World Bank Research Observer* 15, no. 2 (2000): 251–264.

Winpenny, J. *Managing Water as an Economic Resource.* London: Routledge, 1994.

———. "Financing Water for All: Report of the World Panel on Financing Water Infrastructure." Geneva: World Water Council/Global Water Partnership/Third World Water Forum, 2003.

Wolff, G., and E. Hallstein. *Beyond Privatization: Restructuring Water Systems to Improve Performance.* Oakland, CA: Pacific Institute, 2005.

Woo-Cummings, M. *The Developmental State.* Ithaca, NY: Cornell University Press, 1999.

Wood, J. "India's Narmada River Dams: Sardar Sarovar under Siege." *Asian Survey* 33, no. 10 (1993): 968–984.

Woodcock, J. "Indonesia: Enabling Water Utilities to Serve the Urban Poor." Jakarta: World Bank East-Asia Infrastructure Department and Indonesia Country Program, 2006.

World Bank. "Appraisal of the Jakarta Urban Development Project in Indonesia." World Bank, Transportation and Urban Projects Department report no. 475a–IND. Jakarta: World Bank, 1974.

———. "OED Annual Review of Evaluation Results." Washington, DC: World Bank, Operations and Evaluation Department, 1989a.

———. "Developing the Private Sector: A Challenge for the World Bank Group." Washington, DC: World Bank, 1989b.

———. "Office Memorandum: Privatization of the Water Sector in the UK and France." Washington, DC: World Bank/International Finance Corporation, 1991a.

———. "Indonesia Public Expenditure for Infrastructure Development: Background Paper on Urban Development Expenditures." Washington DC: World Bank, 1991b.

———. "Water Supply and Sanitation Projects: The Bank's Experience: 1967–1989." Washington, DC: World Bank Operations and Evaluation Department, 1992a.

———. "Effective Implementation: Key to Development Impact. Report of the Portfolio Management Task Force." Washington, DC: World Bank, 1992b.

——. *Water Resources Management: Policy Paper.* Washington, DC: World Bank, 1993.

——. *World Development Report—1994: Infrastructure for Development.* New York: Oxford University Press, 1994.

——. *Toolkits for Private Participation in Water and Sanitation.* Washington, DC: World Bank, 1997a.

——. "World Development Report 1997. The State in a Changing World." New York: Oxford University Press, 1997b.

——. "Second Jabotabek Urban Development Project: Implementation Completion Report Number 18049, Republic of Indonesia." Washington, DC: World Bank Operations Evaluation Department, 1998.

——. "Water Sector Adjustment Loan: Sectoral Environmental Assessment." E-267. Washington, DC: World Bank and Jakarta, Government of Indonesia, Inter-Agency Task Force on Water Sector Policy Reform, 1999.

——. "Global Development Finance 2000." Washington, DC: World Bank, 2000.

——. "Private Sector Development Strategy: Directions for the World Bank Group." Washington, DC: World Bank, 2002.

——. *Water Resources Sector Strategy: Strategic Direction for World Bank Engagement.* Washington, DC: World Bank, 2003a.

——. *Averting an Infrastructure Crisis: A Framework for Policy and Action.* Jakarta: World Bank, 2003b.

——. "Indonesia: Averting an Infrastructure Crisis; A Framework for Policy and Action." Jakarta: World Bank East-Asia Infrastructure Department and Indonesia Country Program, 2004.

——. "Infrastructure Development: The Roles of the Public and Private Sectors; World Bank Group's Approach to Supporting Investments in Infrastructure. World Bank Guidance Note." Washington, DC: World Bank, 2005.

——. "Approaches to Private Participation in Water Services: A Toolkit." Washington, DC: Public Private Infrastructure Advisory Facility and World Bank, 2006a.

——. *World Development Report—2006: Equity and Development.* New York: Oxford University Press, 2006b.

World Bank and IDA. "Annual Report." Washington, DC: World Bank and International Development Association, 1971.

World Bank/IMF. "Implementing the World Bank Group Infrastructure Action Plan (with Special Emphasis on Follow-up on the Recommendations of the World Panel on Financing Water Infrastructure)." DC2003–0015. Washington, DC: World Bank and International Monetary Fund, Development Committee, 2003.

World Resources Institute. "Seeding the Sustainable Economy: State of the World Report." Washington, DC: World Resources Institute, 2008.

Worster, D. *Rivers of Empire: Water, Aridity, and the Growth of the American West.* Oxford: Oxford University Press, 1985.

WWC. "Official Development Assistance for Water, 1990–2004: Figures and Trends." Marseilles: World Water Council, 2005.

Xenos, N. *Scarcity and Modernity.* London: Routledge, 1989.

Yavari, L. "Management of Basic Drinking Water and Sanitation Services by a Co-operative in Bolivia." In *Reclaiming Public Water: Achievements, Struggles and*

Visions from around the World, edited by B. Balanyá, B. Brennan, O. Hoedeman, S. Kishimoto, and P. Terhorst, 37–44. Amsterdam: Transnational Institute and Corporate Europe Observatory, 2005.

Yayasan Dian Desa. "Monitoring and Evaluation of Public Hydrants and Water Terminals in North Jakarta." Yogyakarta, 1990.

Young, I.M. *Justice and the Politics of Difference.* Princeton, NJ: Princeton University Press, 1990.

Younger, P. "Pro-Poor Water Technologies Working Both Ways: Lessons from a Two-Way, North-South Interchange." *Geoforum* 38, no. 5 (2007): 828–840.

Zaidi, A. *From Lane to City: The Impact of the Orangi Pilot Project's Low Cost Sanitation Model.* London: WaterAid, 2001.

Zwarteveen, M. "Water: From Basic Need to Commodity; A Discussion on Gender and Water Rights in the Context of Irrigation." *World Development* 25, no. 8 (1997): 1335–1349.

INDEX

Note: Page numbers in *italics* indicate figures; those with a *t* indicate tables.

activism. *See* protests, antiprivatization
alternatives, to privatization, 8, 11–16,
107, 143–45, 144t, 164, 215, 218–22.
See also community water supply;
ecological governance; human right,
water as; public-public partnerships
Amnesty International, 146
Anti-Privatization Forum (South Africa),
146
Appadurai, Arjun, 24, 47, 160
archipelagos, networks versus, 22–27, 26t,
28, 110
Arendt, Hannah, 105
Argentina, 32–33, 151; financial crises of,
92, 96; private-public water suppliers
in, 83; water privatization in, 78, 140t,
142
artisanal water supply. *See* community
water supply
Asian Development Bank (ADB), 59,
243n6
Athens, 191, 200
Australia, 190

Bandung Conference (1955), 117
Bangladesh, 46–47
Barlow, Maude, 146
Bechtel Corporation, 2, 166–68
Beck, Ulrich, 195–96
Benn, Hilary, 80
Berg report, 69
biopolitics, 190–212, 221–22
Biwater Corporation, 78–81
Blue October (organization), 137, 249n28
Blue Planet Project, 136–37, 249n30
Boelens, Rutgerd, 40, 208
Bolivia: Cochabamba Declaration and,
249n28, 249n30; irrigation water law,

in, 173–76; private sector contract
cancellations by, 97; privatization
protests in, 140t, 142, 165–69, 173–76,
206; water as human right in, 146, 175;
water cooperatives in, 178, 254n57
Bond, Patrick, 131, 223, 258n9
Boston, 32, 83
Brazil, 213–14, 222, 225; cholera in,
213; concession contracts by, 94, *95*;
participatory budgeting in, 144t, 171t;
privatization protests in, 140t
bribery. *See* corruption
budgeting, 62, 73–74; participatory, 144t,
171t, 213
Buenos Aires, 32–33, 78, 83, 140t, 151.
See also Argentina
Buky report, 68–69
Bustamante, Rocío, 206, 207, 254n37

Caffentzis, George, 172
Cambodia, 214
Camdessus report, 242n95
Canada, 178; community watershed
boards in, 144t, 171t; market environmentalism in, 37t; privatization protests
in, 96, 140t; water as human right in,
136
Canadians, Council of, x, 146, 250n32
capitalism: "natural," 194–95; and
nature, 5, 138–39, 198; Polanyi on,
196; second contradiction of, 194–95;
"vulture," 23, 49, 131. *See also*
Marxism; privatization
Cato Institute, 150
Cayman Islands, 166
Chatterjee, Partha, 24, 29, 30, 105, 160;
on civil/political society, 47–48, 185; on
exclusion, 170; on governance, 50